The Plain English Guide

The **Plain English** Guide

Martin Cutts

Oxford · New York
OXFORD UNIVERSITY PRESS

Oxford University Press, Great Clarendon Street, Oxford OX2 6DP

Oxford New York
Athens Auckland Bangkok Bogota Bombay
Buenos Aires Calcutta Cape Town Dar es Salaam
Delhi Florence Hong Kong Istanbul Karachi
Kuala Lumpur Madras Madrid Melbourne
Mexico City Nairobi Paris Singapore
Taipei Tokyo Toronto

and associated companies in
Berlin Ibadan

Oxford is a trade mark of Oxford University Press

First published in paperback 1996

British Library Cataloguing in Publication Data
Data available

Library of Congress Cataloging in Publication Data
Data available
ISBN 0-19-860049-6

10 9 8 7 6 5 4 3 2

Printed in Great Britain
by Mackays PLC
Chatham, Kent

Contents

Starting points *1*
Summary of guidelines *9*

1 Writing shorter sentences . . . or chopping up snakes 11
2 Preferring plain words 19
3 Writing tight 40
4 Favouring the active voice 48
5 Using vigorous verbs 56
6 Using vertical lists 61
7 Negative to positive 67
8 Cross-references, cross readers 69
9 Clearly non-sexist 71
10 Sound starts and excellent endings 75
11 Using good punctuation 80
12 Seven writing myths explored and exploded 94
13 Conquering grammarphobia 99
14 Planning effectively 102
15 Using reader-centred structure 108
16 Using alternatives to words, words, words 118
17 Management of colleagues' writing 124
18 Writing better instructions 132
19 Lucid legal language 140
20 Clear layout 148

Sources and notes *160*
Index *163*

TO IVOR AND JOAN CUTTS,

for 40 years of support and encouragement

Acknowledgements

My thanks go to those who commented on the manuscript: Nick Moore, Katie Shipsides, Tony Mercer, Ivor Cutts and Kevin Cutts. Throughout the writing of the book, Ingrid Rampley gave constant help and support.

I am also grateful to Monica Sowash, who conducted the focus group interviews; and to Sidney Greenbaum and Ni Yibin, who helped me gain access to the Survey of English Usage.

Extracts from a debt advice booklet in chapter 20 are reproduced by permission of The Birmingham Settlement National Debtline, the copyright owners.

Starting points

Today's post has just arrived, bringing a slip of paper from an insurance company. No surprise in that, as I'm expecting an acknowledgment of a payment I've sent them. The surprise is that it's written like this:

> **The said aggregate further single premium shall be apportioned equally among the existing Policies and consequently in relation to each such Policy the Further Minimum Sum Assured secured by the part of the said aggregate further single premium apportioned thereto shall be a sum equal to the aggregate Further Minimum Sum Assured specified in the Schedule divided by the total number of the existing Policies the Further Participating Sum Assured so secured shall be a sum equal to the aggregate Further Participating Sum Assured so specified divided by the total number of the existing Policies and the amount of the further single premium paid under each of the existing Policies shall be a sum equal to the further aggregate single premium so specified divided by the total number of the existing Policies**

Signed by the company's president and by its managing director, this is a single unpunctuated sentence of 132 words – not even a full stop at the end. A strange way indeed of keeping the customers happy and encouraging them to send more money.

The mail also includes a letter from the manager of the city post office. It begins:

> **I again return your application for renewal of road fund licence and reitterate the correct is amount is £130.**
>
> **I have enclosed a photocopy of your original envelope you will notice it it clearly datestamped 2nd December, some two days after Budget day.**

To make four errors in 43 words doesn't prove the organization is about to collapse, but gives an impression of incompetence. Customers could be forgiven for suspecting the judgement of any writer so careless.

This book is about these two pieces of writing and about the millions of others – many of them, thankfully, much more lucid and well expressed – that are produced every day at work.

The book is not about the writing of novels, plays, poetry or newspapers. It focuses on 'essential information', a loose description that includes business and government letters and reports; consumer contracts; product instructions; leaflets and forms on tax, health, welfare and legal rights; and rules, regulations and laws. These things may seem humdrum compared to the mighty works of literature, but they help to oil the engines of industry, commerce and administration. They provide information which, if misunderstood or half understood, disadvantages people, oppresses them or – at the least – wastes their time and money. It is important to write these documents well.

Guidance not rules

The book has one main aim: to help you write and set out essential information clearly. It does this by suggesting and examining 20 guidelines often used by professional editors.

I say guidelines, not rules. There is a guideline to aim for an average sentence length of 15–20 words in a document, but there is no rule prohibiting sentences of more than 20 words. There is a guideline to use words your readers are likely to understand, but there is no rule prohibiting technical terms.

So this is not meant to be a simple recipe book or a set of quick fixes. Writing will still be hard work – perhaps the hardest work you do – but the guidelines should make it easier and help you produce results that are more readily understood.

Like the two pieces of writing which this chapter began with, the examples used in the book are authentic apart from some of the obvious fictions I have created in chapter 15. I have changed words only to protect authors' anonymity or to add enough context to help you make sense of the examples.

For some of the examples, I have provided rewrites. These are not meant to be the sole or perfect solutions. There are many ways of saying the same thing; if you can improve on my versions, so much the better.

No writing can truly be regarded as clearer or better, however, until users' performance proves it. Several times I mention the value of testing high-use documents with likely users, to see whether they can follow the information and act upon it. I have taken my own advice by testing the clarity of some of the pairs of 'before' and 'after' examples shown. A focus group of 35 people rated the examples for clarity by giving them a score out of 20. The focus group included police officers, firefighters, library staff, unemployed people, teachers, pensioners, booksellers, gardeners – and one person who described herself as a dogsbody. Though focus groups are not statistically representative of the population, they give a useful snapshot of readers' perceptions.

In general, the focus group preferred the versions that took account of the guidelines, rating them significantly clearer than the originals. This does not prove that the performance of group members in a practical test would have been better, but it is certainly a strong indicator. The results, shown at appropriate points in the text, enable me to recommend the guidelines to you with even greater confidence.

What is plain English?

The guidelines will help you create plain English documents. But what is really meant by plain English? Is it anything more than a slogan used by campaigners to publicize themselves and their favourite cause, and by businesses selling editing and document design services?

Undoubtedly, plain English is a woolly term. As no formula can genuinely measure the plainness of a document, I would rather describe plain English than define it. In my view, plain English refers to:

> **The writing and setting out of essential information in a way that gives a co-operative, motivated person a good chance of understanding the document at first reading, and in the same sense that the writer meant it to be understood.**

This means pitching the language at a level of sophistication that suits the readers and using appropriate structure and layout to help them navigate through the document. It does not mean always using simple words at the expense of the most accurate words or writing whole documents in kindergarten language – even if, as some adult literacy surveys claim, some seven million adults in the UK and about 70 million adults in the US cannot read and write competently.

'Plain' has connotations of honesty, or should have. Essential information should not lie or tell half-truths, especially as its providers are often socially or financially dominant. For example, insurance policies should not hide management charges beneath a mass of detail and doublespeak. A health advisory booklet should not withhold facts that happen to be unpalatable to political, religious or minority groups. Products claiming green credentials should not use duplicitous labels like 'recyclable', since almost anything can be recycled, or 'harvested from sustainable forests', since all forests are sustainable and what matters is whether they are indeed sustained, and how.

Plain English is not an absolute: what is plain to an audience of scientists or philosophers may be obscure to everyone else. And because of variations in usage across the English-speaking world, what is plain in Manchester may be obscure in Madras or Maine. Similarly, what is plain today may be obscure a hundred years from now because patterns of usage, readers' prior knowledge, and readers' expectations will all alter over time.

Where did plain English spring from?

Pleas for plain English have been around for a long time. In the fourteenth century Chaucer had one of his characters demand:

> **Speketh so pleyne at this time, I yow preye**
> **That we may understonde what ye seye.**

In 1500, *Chaucer's Dreme* refers to a story 'which ye shalle here in pleyne Englische'. The few remaining writings of William Tyndale do not reveal whether he saw himself as plain Englishing the Latin Bible in 1525. But his influential translation, for which he was executed, certainly used the direct and pungent voice of the common people of his day. And Tyndale seems to have been keen on the modern plain English principle of keeping the readers in mind, if his growl to one opponent is anything to go by: 'If God spare my life, ere many years I will cause a boy that driveth the plough shall know more of the scripture than thou dost.'

In 1550, after only three years on the throne of England, Edward VI had become so exasperated with the law that he remarked:

> **I would wish that the superfluous and tedious statutes were made more plain and short, to the intent that men might better understand them.**

In the same century some writers wanted to halt the influx of what they called inkhorn terms such as 'revoluting', 'ingent affabilitie' and 'magnifical dexteritie', which were meant to broaden what could be said in English and make it sound more grand. Some inkhorn terms like 'defunct' and 'inflate' survive to this day – perhaps even as plain words – because people have found them useful.

In 1604 the first-ever dictionary in English sought to explain:

> **hard vsuall English wordes, borrowed from the Hebrew, Greeke, Latine, or French. &c. With the interpretation thereof by plaine English words, gathered for the benefit & helpe of Ladies, Gentlewomen, or any other vnskilfull persons.**

Patronizing though this seems today, the dictionary was meant to help people – especially women – whose lack of access to private tutors and grammar schools left them unable to understand the heavily latinized English that was then so fashionable among the high and powerful.

From the seventeenth century, Protestants, especially Quakers, tended to favour a simple style in their writing and speaking. They called it plain language. In nineteenth-century England, others sought a weird kind of linguistic purity by trying to replace Latin-derived words with those of a Saxon look. William Barnes, for example, preferred 'leechcraft' to 'medicine', 'speechlore' to 'grammar', and 'swanling' to 'cygnet'. His terms 'foreword' (instead of 'preface') and 'handbook' (instead of 'manual') are today as popular as their alternatives. The English barrister George Coode began trying to get legal sentences written more clearly. He said, rather too optimistically:

> **Nothing more is required than that instead of an accidental and incongruous style, the common popular structure of plain English be resorted to.**

The philosopher Jeremy Bentham called for laws to be divided up into sections (this was being done in France), demanded shorter sentences (one law included a 13-page sentence) and suggested that lawyers should 'speak intelligibly to whom you speak'.

In England in the 1920s, C K Ogden and I A Richards devised Basic English. Its core was a vocabulary of 850 words which, in various combinations and using a narrow range of sentence structures, could, they believed, say everything that needed to be said. Their three aims were that Basic should be an international language, an introduction to full standard English for foreigners, and a kind of plain language for use in science, commerce and government. Basic was supported by Winston Churchill, the British prime minister, and by Theodore Roosevelt, the US president. Though influential to this day in the teaching of English as a foreign language, Basic became bogged down in academic controversy and eventually fizzled out as a force for plain language in the 1950s.

In that decade a landmark for plain English was erected when the Treasury commissioned Ernest Gowers, a top civil servant, to write a book encouraging plain writing in the civil service and elsewhere. This eventually became *The Complete Plain Words,* still widely read in updated editions. Unfortunately the book continues to exempt many forms of legal drafting from its advice, a loophole that lawyers plead in mitigation of their often wayward writing habits.

Modern developments

In the 1970s the seeds of a plain English revolution were sown. In the US and the UK, consumer groups used the mass media to publicize and ridicule examples of obscurity in legal documents and government forms, calling for plain language or plain English. The American Council of Teachers of English formed a Public Doublespeak Committee to draw attention to the use of deceptive language by public figures, and gave a George Orwell Award for honesty and clarity in public language. In Australia the first car insurance policy that could reasonably be called 'plain' was issued. In the UK there was even a whiff of book-burning in 1979 as campaigners for plain English shredded unclear government forms outside the Houses of Parliament.

In the US, President Carter signed executive order 12044 requiring regulations to be written in plain English, though this was repealed by his successor in 1981. Some states, like New Jersey, passed laws requiring consumer contracts to satisfy certain standards of clarity in their language and layout.

In 1982 the British government issued a White Paper (a policy statement) ordering departments for the first time to count their forms, abolish unnecessary ones, clarify the rest, and report their progress annually to the prime minister. By the end of the 1980s, under this onslaught, it was difficult to find a truly atrocious central government form in the UK. Local government has also been influenced. Some town halls have set up plain English committees of elected members and officials who vet forms and leaflets for clarity before they are printed. One puts a logo on its approved documents saying 'Plain English by Derby City Council'. Similar accreditation schemes are also sold commercially.

In 1984 the Australian government adopted a plain language policy for its public documents and slowly this is being extended to the language of the law itself. Law-writers in Canberra now take advice from their own 'Plain English Manual'. In Queensland the Industrial Relations Reform Bill 1993 even legislates for clarity – tribunal decisions must be 'expressed in plain English' and 'structured in a way that is as easy to understand as the subject matter allows'. There is a strong political push behind this. Politicians not only smell a vote or two in plain English, as it removes another remnant of Australia's colonial past, but they see it as a way of cutting the cost of government and compliance with regulations.

European law has given a powerful boost to the use of plain language in certain standard-form consumer contracts. EC Council directive 93/13 requires unfair terms to be removed, and:

> **In the case of contracts where all or certain terms offered to the consumer are in writing, these terms must always be drafted in plain, intelligible language. Where there is doubt about the meaning of a term, the interpretation most favourable to the consumer shall prevail.**

This has persuaded many previously stick-in-the-mud companies to write their contracts more plainly.

Does plain English work?

Research shows that documents carefully crafted in plain English can improve readers' comprehension. In an American study of instructions given by word of mouth to jurors, the plain versions improved comprehension by 31 per cent, from 45 per cent to 59 per cent. In a further study the same instructions were given in both speech and writing. Jurors understood the plain versions 'almost fully', said the researchers. In a study to test understanding of medical consent forms in the US, readers of the original form could answer correctly only 2.36 questions out of 5. Using the plainer form they could answer 4.52 questions, a 91 per cent improvement; they also took much less time to answer. In the UK the Plain Language Commission tested my Clearer Timeshare Act (a law rewritten in plain English) with 90 senior law students. Nine out of 10 preferred the plain version to the real act. Performance also improved:

when answering one key question, 94 per cent got the correct result when working with the rewritten version, while only 48 per cent did so with the real act.

Much of the most convincing research on the benefits of plain English relates to legal documents. And the most telling point of all is that no company that has issued a plain English insurance policy, pension contract or bank guarantee has ever reverted to a traditional legalistic style of wording.

The point of plain English

What has motivated me and many others to work for the plain English cause is that clearer documents can improve people's access to benefits and services, justice and a fair deal. If people understand what they are asked to read and sign, they can make better choices and know exactly what they are letting themselves in for. They might even see more clearly what business and government are up to. Plain language should, I believe, become an accepted part of plain dealing between consumers and business, and between citizens and the State. I hope this book will help it to do so.

Summary of guidelines

Most of the chapters begin with a guideline which is then expanded and discussed. The 20 guidelines are set out below. Numbers correspond to chapter numbers.

Style and grammar

1 Over the whole document, make the average sentence length 15 to 20 words.

2 Use words your readers are likely to understand.

3 Use only as many words as you really need.

4 Prefer the active voice unless there's a good reason for using the passive.

5 Use the clearest, crispest, liveliest verb to express your thoughts.

6 Use vertical lists to break up complicated text.

7 Put your points positively when you can.

8 Reduce cross-references to the minimum.

9 Try to avoid sexist usage.

10 In letters, avoid fusty first sentences and formula finishes.

11 Put accurate punctuation at the heart of your writing.

12 Avoid being enslaved by writing myths.

13 You can be a good writer without learning hundreds of grammatical terms.

Preparing and planning

14 Plan before you write.

Organizing the information

15 Organize your material in a way that helps readers to grasp the important information early and to navigate through the document easily.

16 Consider different ways of setting out your information.

Management of writing

17 Manage colleagues' writing carefully and considerately to boost their morale and effectiveness.

Plain English for specific purposes: instructions and legal documents

18 Devote special effort to producing lucid and well-organized instructions.

19 Apply plain English techniques to legal documents such as insurance policies, car-hire agreements, laws and wills.

Layout

20 Use clear layout to present your plain words in an easily accessible way.

1 Writing shorter sentences . . . or chopping up snakes

Guideline: *Over the whole document, make the average sentence length 15 to 20 words.*

More people fear snakes than fear full stops, which could explain why they recoil when a long sentence comes hissing across the page. Here's one from an accountant's letter to his self-employed client:

> **Our annual bill for services (which unfortunately from your viewpoint has to increase to some degree in line with the rapid expansion of your business activities) in preparing the accounts and dealing with tax (please note there will be higher-rate tax assessments for us to deal with on this level of profit, which is the most advantageous time to invest in your personal pension fund, unless of course changes are made in the Chancellor's Budget Statement) and general matters arising, is enclosed herewith for your kind attention.**

At one level, the sentence is easy to read – all the words are clear enough for a literate client. What makes it hard work is its length and muddle, with asides and additions tagged on as they sprang into the writer's mind.

Muddle is more likely in a long sentence unless the construction is simple and well organized. Instead of making one main point – and perhaps one related subsidiary point – a long sentence force-feeds the reader with point after point after point. This demands more concentration and short-term memory, leading to overload if the topic is complicated. If you want to get your recommendation agreed by a board of directors, it is unwise to waste their time disentangling a 68-word sentence:

> **Although Mr Smith's requested increase in borrowing facility is substantial and the Smith Group has not yet returned to acceptable profitability, in fact because of dividends paid there was a deficit of £10 million transferred to revenue reserves, profits at the trading level have increased with considerable opportunity to advance**

further as the recession eases and consumer confidence improves, therefore an increased limit to £12.5 million is recommended.

But what length of sentence is too long? Ignore advice that prescribes an upper limit, though if you regularly exceed 40 words you will certainly weary your readers. Better to aim for an average of 15 to 20 words throughout. The key word is *average*, so not all sentences need to be in this range; there should be plenty of variety. While an occasional short sentence will punch home an important point effectively, too many will make your writing dull and staccato. There is no lower limit: a sentence could be just one word, such as 'Why?', or two words, such as 'I disagree'.

One reason for keeping to a 15–20 word average is that people are used to it. In the mid-1960s, a million words of published US writing were analysed. The average sentence length was 19. Scores for particular types of document included:

- Miscellaneous: government 25
- Learned and scientific 24
- Press: reviews 22
- Press: reports 21
- Humour 18
- Fiction: romance/love 14
- Fiction: science/detective 13

It's notable that government and scientific documents, perhaps the only material that could be classed as essential information, came top.

If you find yourself writing long sentences or having to edit them for other people, there are six main ways of clarifying them:

- Split and disconnect.
- Split and connect.
- Say less.
- Use a list.
- Cut verbiage.
- Bin the sentence and start again.

Let's examine each of these.

Split and disconnect

Full stops enable readers to digest your latest point and prepare for the next. This sentence from a local government report makes good sense but, at 57 words, is too long for busy people to grasp at first reading:

> **I understand that some doctors making night calls have been attacked in recent months on the expectation that they were carrying drugs and their caution when visiting certain areas in the south of the city has been very exacting and has even included telephoning the address to be visited from their car when they arrive outside the house.**

Look for the main break in the sense. It comes in the third line after 'drugs', where the writer starts talking about caution. So you could split it there, delete 'and', and produce this:

> **I understand that some doctors making night calls have been attacked in recent months on the expectation that they were carrying drugs. Their caution when visiting certain areas in the south of the city has been very exacting and has even included telephoning the address to be visited from their car when they arrive outside the house.**

That may be enough. But you could go further and split it again after 'exacting', producing this:

> **I understand that some doctors making night calls have been attacked in recent months on the expectation that they were carrying drugs. Their caution when visiting certain areas in the south of the city has been very exacting. It has even included telephoning the address to be visited from their car when they arrive outside the house.**

Now the paragraph can be readily grasped at first reading.

Split and connect

This means putting in a full stop and restarting the sentence with a connecting word like 'however', 'but', 'so', 'also', 'yet' or 'further'. The technique would be helpful in this example from a lawyer's letter:

> **Whilst it is expected by the donor's family that the present arrangement for caring for the donor will continue for the rest of her life, should it at any stage become necessary to transfer the donor once more into a nursing institution, the donor's family envisages that the second-floor flat will be sold and the donor's share in the proceeds used to provide any additional income necessary to ensure her continued well-being.**

That's a 73-word sentence – not enormous, perhaps, but certainly a candidate for splitting. The obvious break point is after 'life' in the third line:

> **Whilst it is expected by the donor's family that the present arrangement for caring for the donor will continue for the rest of her life.**

But this fails because the sentence is unfinished – nothing complete has yet been said. Striking out 'Whilst', however, creates a complete sentence and the next can begin with 'but' (see chapter 12 if that worries you):

> **~~Whilst~~ It is expected by the donor's family that the present arrangement for caring for the donor will continue for the rest of her life. But should it at any stage become necessary to transfer the donor once more into a nursing institution, the donor's family envisages that the second-floor flat will be sold and the donor's share in the proceeds used to provide any additional income necessary to ensure her continued well-being.**

Other improvements are possible. At the start, it would be simpler to say 'The donor's family expects . . .'. The second sentence could be full-stopped at 'sold', with a new one beginning 'The donor's share in the proceeds would then be used . . .'.

Say less

Sometimes a sentence is lengthened by needless repetition, as at the start of this lawyer's letter:

> **Dear Sirs**
> **Trial of John Smith and James Jackson**
> **Trade Descriptions Act 1968, Manchester Crown Court, 10.30am, Tuesday 7 June 199-**
> **The above defendants are to be tried at the Manchester Crown Court on Tuesday 7 June 199- at 10.30am for several offences under the Trade Descriptions Act 1968 concerning the supply of motor vehicles to which false trade descriptions had been applied.**

As the first sentence repeats most of the heading, it could say:

> **Dear Sirs**
> **Trial of John Smith and James Jackson**
> **Trade Descriptions Act 1968, Manchester Crown Court, 10.30am, Tuesday 7 June 199-**
> **The above defendants are to be tried for several offences concerning**

the supply of motor vehicles to which false trade descriptions had been applied.

That saves 18 words and preserves the meaning. It also lets the heading do the job it's supposed to do.

Use a list

Vertical lists break up long sentences into manageable chunks. They are particularly useful when describing a procedure, as in this safety message to hospital staff about a machine for keeping babies warm:

The attachment of the warmer support-bearing assembly system must be checked to ensure that it is adequately lubricated, its securing screws are tight and that the warmer head can be easily repositioned without the support bearing sticking.

Not a hard sentence for people in the trade – and only 38 words long – but easier to follow in a vertical list:

The attachment of the warmer support-bearing assembly system must be checked to ensure that:
(a) it is adequately lubricated;
(b) its securing screws are tight;
(c) the warmer head can be easily repositioned without the support bearing sticking.

There's more on lists in chapter 6.

Cut verbiage

Sometimes the key ideas in a long sentence are buried under verbiage:

The organizers of the event should try to achieve greater safety both from the point of view of ensuring that the bonfire itself does not contain any unacceptably dangerous materials such as aerosol cans or discarded foam furniture and from the point of view of ensuring the letting-off of fireworks in the designated area, with easily identifiable wardens to be available during the event to prevent people indiscriminately letting off fireworks, to the possible danger of people attending the event.

The redundant words include 'from the point of view of' (twice), 'itself', 'unacceptably' (what are 'acceptably' dangerous materials?), 'discarded' (they wouldn't be on the bonfire otherwise), 'during the event' and 'to the possible danger of people attending the event' (redundant because crowd safety is what the whole sentence is about). With verbiage crossed through and insertions underlined, the sentence could be:

> **The organizers of the event should try to achieve greater safety ~~both~~ by ~~from the point of view of~~ ensuring that the bonfire ~~itself~~ does not contain any ~~unacceptably~~ dangerous materials such as aerosol cans or ~~discarded~~ foam furniture and ~~from the point of view of~~ by ensuring the letting-off of fireworks in the designated area, with easily identifiable wardens to be available during the event to prevent people indiscriminately letting off fireworks~~, to the possible danger of people attending the event~~.**

Then, with a few minor modifications and a full stop after 'area', the sentence becomes:

> **The organizers of the event should try to achieve greater safety by ensuring that the bonfire does not contain any dangerous materials such as aerosol cans or foam furniture, and that fireworks are let off only in the designated area. Easily identifiable wardens should be present to prevent people letting off fireworks indiscriminately.**

The first sentence could also be converted into a list, using the first 'that' as a pivot and deleting the second 'that':

> **The organizers of the event should try to achieve greater safety by ensuring that:**
>
> - **the bonfire does not contain any dangerous materials such as aerosol cans or foam furniture;**
> - **fireworks are let off only in the designated area.**

There's more on cutting verbiage in chapter 3.

Bin the sentence and start again

When there's no hope of untangling a sentence by other means, all you can do is discard it and rewrite. Here is that accountant's sentence from the top of the chapter, sadly unimproved since we left it:

> **1/ Our annual bill for services (which unfortunately from your viewpoint has to increase to some degree in line with the rapid expansion of your business activities) in preparing the accounts and dealing with tax (please note there will be higher-rate tax assess-**

ments for us to deal with on this level of profit, which is the most advantageous time to invest in your personal pension fund, unless of course changes are made in the Chancellor's Budget Statement) and general matters arising, is enclosed herewith for your kind attention.

It's no good just putting in full stops because the new sentences won't make any sense, and using a vertical list doesn't seem feasible. Nor does it help to make fewer points – it's all useful detail, though verbiage like 'from your viewpoint', 'general matters arising', 'herewith' and 'kind' could all be cut. So we need to start again and plan out the main points, which are:

(a) Here is our annual bill for services.

(b) We're charging more than last year for two reasons: your business has grown rapidly, and we'll have to work out a higher-rate tax assessment as you've made a lot more profit.

(c) Now is a good time to pay into your pension fund as you'll get higher-rate tax relief on your contributions, unless the rules alter in the Chancellor's budget.

This isn't the best sequence of points – the good news ought to come first. So, with the points in c-a-b order and written as sentences, the result could be:

2/ Now is a good time for you to pay into your pension fund as you will get higher-rate tax relief on your contributions – unless the Chancellor's budget changes the rules.

I have enclosed our annual bill for services. Unfortunately it is higher than last year. This is because your business has grown rapidly and, since your profits are much greater, I will need to calculate a higher-rate tax assessment.

Four sentences – and readily understood at first reading. If you prefer a more informal style, you could put in some contractions (such as *I've*, *You'll*):

3/ Now is a good time for you to pay into your pension fund as you'll get higher-rate tax relief on your contributions – unless the Chancellor's budget changes the rules.

I've enclosed our annual bill for services. Unfortunately it's higher than last year. This is because your business has grown rapidly and, since your profits are much greater, I'll need to calculate a higher-rate tax assessment.

The last paragraph could also be written as a vertical list:

4/ Now is a good time for you to pay into your pension fund as you'll get higher-rate tax relief on your contributions – unless the Chancellor's budget changes the rules.

I have enclosed our annual bill for services. Unfortunately it is higher than last year. This is because:

- **your business has grown rapidly; and**
- **I will have to work out a higher-rate tax assessment as your profits are much greater.**

A 35-strong focus group (see page 3) assessed three of the four versions for clarity. They rated Version 4 the clearest, with an average score of 18 points from a possible 20. It was the first preference of 26 people. Version 2 got an average score of 14 but only three first preferences. Version 1, the original, scored only 5 points – seven respondents gave it no points at all. These results suggest that sentence length and layout significantly affect readers' perception of clarity.

Why bother?

Few people think any less of accountants who occasionally write a bad sentence; after all, they're mainly in the figures business. But members of any profession who write snake-like sentences invite readers to lose confidence in them and to take their business elsewhere. As the focus group showed, people do notice the difference between an accountant who writes considerately and one who doesn't. If only for commercial reasons, it is important to control sentence length carefully. Fortunately this is easily done.

2 Preferring plain words

Guideline: *Use words your readers are likely to understand.*

Here is a Secretary of State refusing an assistant's request for a pay rise:

> **Because of the fluctuational predisposition of your position's productive capacity as juxtaposed to government standards, it would be momentarily injudicious to advocate an increment.**

And here is a local government official being obstructive to citizens who wish to display posters in a public library:

> **Your request raises a question as to the provenance and veraciousness of the material, and I must consider individually all posters of a polemic or disputatious nature.**

Both sentences over-dress simple ideas, clothing them in phrases designed to impress not inform. Once, in forelock-tugging times, people were impressed. Now, they smell pomposity and dislike being put to the trouble of translating.

For business writers this may have costly consequences. An accountancy firm wrote this in a business proposal:

> **At present the recessionary cycle is aggravating volumes through your modern manufacturing and order processing environments which provide restricted opportunities for cost reduction through labour adjustments and will remain a key issue.**

What they probably meant was:

> **Output and orders have fallen because of the recession but there is little scope for reducing the workforce.**

Whatever they meant, they didn't win the business.

If a selling point is obscured by gobbledygook it ceases to be a selling point. An aeronautics firm wrote to a prospective customer:

We would anticipate being able to optimize the engine design from an emissions point of view.

In such a haystack of verbiage, the needle of meaning could easily have been missed:

We expect to be able to redesign the engine to reduce emissions.

Especially outmoded is the ceremonial language found in town-twinning proclamations – written by normally down-to-earth city governments – and in royal pronouncements such as this citation on the Queen's Award for Export Achievement:

XYZ Company Limited: Greetings! We being cognizant of the outstanding achievement of the said body as manifested in the furtherance and increase of the Export Trade of our United Kingdom of Great Britain and Northern Ireland, our Channel Islands and our Isle of Man and being desirous of showing Our Royal Favour do hereby confer upon it the Queen's Award for Export Achievement.

Monarchs, like captains of any great corporation, should demand that their staff put clear, if dignified, language into their mouths. Be eloquent, not grandiose.

As English-speaking countries become more informal, many official bodies have promised to use plain language in their dealings with the public. Sir Ernest Gowers' motto, 'Be short, be simple, be human', has begun to take root. When people expect plain words, they are even more surprised to get officialese.

Is there no place for unusual words?

The chief reason for writing at work is to give information in a way that's readily understood – to bridge the gap between what you know and what the reader knows. Generally, plain words do this best. If you want to say:

Before the court case started, the usher asked everyone for silence

those are the plain words you would use. You wouldn't make people struggle with:

Prior to the commencement of the court case, the usher requested all persons present to adopt a mantle of quietude.

Not that there is anything wrong with long or unusual words (and many long words aren't unusual – think of 'immediately' or 'disappointment'). Sometimes an unusual word is exactly right for the job, expressing just what you want to say. Then you should use it and either give an explanation or trust the context to explain; you should not always hope that busy readers will be willing to consult a dictionary, even supposing they own one. In a technical document, there is a place for technical words, which will be plain enough to technical people. But while doctors might readily understand 'cardiac atheroma' and 'pulmonary oedema', a mass audience would get a clearer sense of the ailments from 'furring-up of the heart's arteries' and 'fluid in the lungs'. There is just no point saying that a person is exhibiting xanthochromia and diaphoresis if you can equally well say he is yellow and sweating. Generally, therefore, the place for unusual words is in literature and journalism, where readers are prepared to travel further with a writer's flight of fancy.

Sometimes educated writers are surprised that words they think ordinary are unknown to the person sitting next to them on the train. Take that example about displaying posters in the library:

> **Your request raises a question as to the provenance and veraciousness of the material, and I must consider individually all posters of a polemic or disputatious nature.**

Of the hundreds of people I've shown it to, only a handful have known that 'provenance' means 'origin' or 'source'. That may be regrettable but it sends a message to anyone with a wide vocabulary: don't assume others know all the words you do.

In the queue at a bank cash point I watched a young man trying to decide from his balance slip whether his account held enough money to permit a withdrawal. He read it out to his friend: '£50.23 CR', but neither of them could tell whether CR (short for 'credit') meant he owed money to the bank or vice versa. In the end, they decided he was in debt and left without withdrawing any cash. Wrong decision, of course, but understandable since if you get credit you are putting yourself in debt. Writers who want to be readily understood have a very tough job: they have to guess what words (or symbols) the reader is likely to understand, and write accordingly.

A similar problem arises with a word favoured by lawyers: 'notwithstanding'. When testing people's comprehension of legal documents, I've found that few non-lawyers can explain it – even in context. Alternative plain words are available: 'despite', 'in spite of', 'however' or 'but' can all do the same job. 'Notwithstanding', like 'forsooth', 'peradventure' and 'verily', has virtually disappeared from everyday speech and writing. Removing it and other oddities from legal documents will benefit people, not hammer a nail in the coffin of mass literacy by reducing their exposure to unusual words.

We shouldn't worry if archaic words disappear from time to time. When people no longer understand a word, or if it is not doing a useful job, its day is done and other words will move in to fill the gap. This has happened throughout history. Once, it was considered correct to use 'ye' in such constructions as 'ye believe', 'ye said' and 'gather ye rosebuds while ye may'. Now, if anyone uses 'ye' like this, we think they have stepped out of a time machine. 'You' has ousted its rival and we seem to be none the worse for the change.

The scope for using plain words – and the opposite – is rich in English because we have so many words with the same or similar meanings ('synonyms'). For example, the cauldron which bubbled for centuries with bits of Latin, French, Old Norse and Old English threw up several synonyms for 'start' – 'begin', 'commence', 'initiate', 'institute' and 'originate'. There is room for all these words, but only 'start' and 'begin' are first choices in today's plain English lexicon.

Let's look at two main ways of removing officialese and pompous language:

- Use simpler alternatives.
- Reconstruct the sentences.

Use simpler alternatives

In this section, for the sake of clarity, the officialese and the equivalent plain English are underlined.

A local government department is writing to a tenant who has fallen behind with her rent. In law, the authority doesn't have to rehouse tenants it regards as deliberately homeless:

> **In the event of your being evicted from your dwelling as a result of wilfully failing to pay your rent, the council may take the view that you have rendered yourself intentionally homeless and as such it would not be obliged to offer you alternative permanent housing.**

Using plain words and splitting the sentence, this could become:

> **If you are evicted from your home because you deliberately fail to pay your rent, the council may decide that you have made yourself**

intentionally homeless. If this happens, the council does not need to offer you alternative permanent housing.

Our focus group found this version far clearer, with 31 out of 35 preferring it. They gave it an average clarity mark of 17 out of 20, as against 12 for the original (a remarkable show of tolerance towards officialese).

A hospital wants to resolve the chaos in its car parks by charging parking fees:

If my proposals are accepted, the income from fees would ensure that car parking control could be effected without utilising monies that should be expended on health care.

This becomes:

If my proposals are accepted, the income from fees would ensure that car parking could be controlled without using money that should be spent on health care.

An official is writing to a citizen about a claim for housing benefit:

I am in receipt of information from the citizens advice bureau, which I believe is acting on your behalf, with regard to matters appertaining to your benefit claim. Will you please furnish the bureau with particulars of your savings.

This becomes:

I have received information from the citizens advice bureau, which I believe is acting on your behalf, about your benefit claim. Will you please give the bureau details of your savings.

A firm's conditions of service say:

Holidays will be taken by mutual agreement after the exigencies of the service have been considered.

This becomes:

Holidays will be taken by mutual agreement after the needs of the service have been considered.

Reconstruct the sentences

The technique is to spot the unusual word or phrase and use its plain meaning as an aid in revising the sentence.

A trade union writes to its members:

It behoves management to give details of the planned redundancies, and it is incumbent on all members to participate fully in this dispute.

Sending the obsolete words back to their vaults, this becomes:

Management should [or 'has a duty to'] give details of the planned redundancies, and all members should [or 'must'] participate fully in this dispute.

A treasurer's report explains how many errors are being made when handling payments:

An approximate frequency for the mistakes was given by Mrs Jones as ten a month.

This becomes:

Mrs Jones said that about ten mistakes a month were being made.

J Edgar Hoover, when head of the Federal Bureau of Investigation in the United States, was worried about Picasso visiting the country (he was said to have left-wing leanings). Hoover wrote to a special agent in Paris:

In the event information concerning Picasso comes to your attention, it should be furnished to the Bureau in view of the possibility that he may attempt to come to the United States.

He might have said:

If you get any information about Picasso, please tell the Bureau in case he tries to visit the United States.

A company training officer is explaining a trend towards computer-based training:

The ready availability of computer-based tutorials associated with applications software has become prevalent since the development of Microsoft Windows.

To a fellow professional, none of these words would be hard to understand. But look closely at what is being said: 'The ready availability . . . has become prevalent'. Since 'ready availability' and 'prevalence' are so similar in meaning, the writer was probably trying to say:

Computer-based tutorials associated with applications software have become readily available since the development of Microsoft Windows.

Plain words make the meaning clearer and the sentence crisper.

An example of good plain words

Writing is harder when you lack the freedom to tackle interesting topics like music, marine life and football. Your writing day could be occupied by matters so humdrum they would make a great novelist weep: the rent for a piece of waste land; inefficient plumbing in the office toilets; a small rise in the price of custard.

Faced with a dull topic, there is great skill in creating a clear document that rises above the bland and boring. The following letter from an official to an anxious elderly couple is a model of simplicity. It doesn't give them everything they want – that would be an easy letter to write – but it does explain things reassuringly and fully. No doubt it could still be improved, but here it is, untouched:

Dear Mr and Mrs France

I write concerning an enquiry on your behalf by Councillor Jameson on several matters relating to your home. One of my colleagues will be writing to you about the bath, as I understand you are now having some difficulty with this.

As to the decoration of a further bedroom, I am arranging for the foreman painter to call and measure up in order to do the work. I would expect this to be early in the new year so that the work can be done before the end of March.

At present there are no plans to install central heating in Jasmine Row. None of the properties in Jasmine Row were included in this year's programme. The programme of installations for next year and future years has not yet been drawn up and I cannot say at this stage whether Jasmine Row will be included in it.

It is our policy to install central heating in all our properties in the next few years, and we are steadily working towards this goal. As soon as I have any definite information I will write to you setting out our proposals.

If I can help further, please contact me.

Conquering fear

The main cause of bloated, show-off writing is fear. First there is fear that being clear means being definite and that being definite leaves no room for wriggling. Yet if writers want to hedge, a plain English style lets them.

They can evoke a full range of doubt and uncertainty – and even the possibility of being just plain wrong – with such words as 'may', 'might', 'could', 'should', 'perhaps', 'normally', and 'generally'. Plain English doesn't always mean being unsubtle.

Then there is the fear that if you write simply you will not be thought sufficiently eminent, scientific or literary. This fear was examined in a research study in 1978. Over 1,500 scientists from industry and the academic world were asked their opinion of two short pieces of scientific writing. Both pieces gave exactly the same facts in the same sequence of points, and used the same five technical terms. The only difference was in the style of the non-technical language. The first version used everyday words and short, simply constructed sentences. The second did the opposite, though without going to extremes. Nearly 70 per cent of the scientists preferred the plain version; they also found it 'more stimulating' and 'more interesting'. Three-quarters judged the writer to be more competent as a scientist and to have a better organized mind.

So don't be afraid of plain English. Carefully used, it will reveal your competence far better than the wooden style of so many academic and technical journals.

Plain English lexicon

If you overuse the words in the left-hand column of the table, your writing could be perceived as pompous, officious and long-winded. Not that anyone should forbid you from ever using them, but judicious use of the alternatives will help you to be shorter, simpler and more conversational. The alternatives are not always synonyms, so use them with a proper care for meaning and for the job they have to do in the sentence.

Official terms	***Plainer alternatives***
accede	agree, grant, allow
accordingly	so
accustomed to	used to
acquaint yourself	find out, read
additional	more, extra
address (sense 'consider')	tackle, deal with, consider
advices	information, instructions
advise (sense 'inform')	inform, tell

Official terms	*Plainer alternatives*
aforementioned, aforesaid	[omit or be specific]
aggregate (noun)	total
alleviate	ease, reduce, lessen
apprise	inform, tell
as a consequence of	because
ascertain	find out
assist, assistance	help
attain	reach
attribute (verb)	earmark
calculate	work out, decide
category	group
cease	stop, end
cognizant of	aware of, know about
commence	start, begin
component	part
concept	idea
concerning	about
consequently	so
constitute	make up, form
construe	interpret
corroboration	evidence, proof, support
deduct	take away, take off, subtract
deem	treat as, consider
defer	put off, postpone
desist	stop
despatch	send
despite the fact that	although, despite
determine	decide
disburse	pay
discharge (verb)	pay off, settle
disconnect	cut off
discontinue	stop, end

Official terms	*Plainer alternatives*
due to the fact that	as, because
dwelling, domicile	home, property
elect; election	choose; choice
emanate from	come from, stem from
endeavour	try, attempt
entitlement	right
envisage	expect, imagine
equitable	fair
erroneous	wrong, mistaken
establish	set up, create, form
eventuate	result, occur, happen
expedite	hasten, speed up
expenditure	spending
expiration	end
facilitate	help
failure to	if you do not
for the duration of	during, while
for the purpose of	to
forward (verb)	send, give
furnish	give, provide
henceforth	from now on
heretofore	until now
herewith	with this
hitherto	until now
hypothecate	earmark
if this is not the case	if not
if this is the case	if so
impart	give, pass on, tell, inform
implement (verb)	carry out, do
in accordance with	in line with
inasmuch as	because, in that
incidence	rate of occurrence, how often

Official terms	*Plainer alternatives*
in conjunction with	with
increment	step, increase
indebtedness	debt
initiate	begin, start
in lieu of	instead of
in order to	to
in receipt of	get, have, receive
in regard to	about, concerning, on
insofar as	as far as
institute (verb)	begin, start
interim (noun)	meantime
in the event of	if, when
in the eventuality of	if, when
in view of the fact that	as, because
manner	way
monies	money, amounts of money
necessitate	need, have to, require
nevertheless	even so, however, yet
nonetheless	even so, however, yet
not less than [ten]	at least [ten]
not more than [ten]	[ten] or less, [ten] or fewer
notwithstanding	even if, despite, still, yet, but
obtain	get, receive
other than	except
particulars	details, facts
persons	people
peruse	read, study
polemical	controversial
principal (adjective)	main, chief
prioritize	rank
prior to	before
provenance	source, origin

Official terms	*Plainer alternatives*
provisions [of a law, policy]	the law, the policy
purchase	buy
purport (verb)	pretend, claim, profess
pursuant to	under
reduction	cut
regarding	about
reimburse	repay
remittance	payment
remuneration	pay, wages, salary
remunerative employment	paid work
render	send, make, give
reside	live
residence	home, property, address
save (co-ordinator)	except
shall [legal obligation]	must
stipulate	state, set, lay down
sufficient	enough
supplementary	extra, more
terminate	end, stop
the law provides that	the law says
thereafter	then, afterwards
timeously	in good time
tranche	slice, portion, share, chunk
utilize	use
verify	check, prove
whensoever	when, whenever
whereby	by which, because of which
whilst	while
wilfully	deliberately
with reference to	about, concerning
with regard to	about, concerning
with respect to	about, for, concerning

Frequency count

It makes sense to use plain words in essential information because people are already familiar with them from other texts, as shown by analysis of a set of 400,000 words of recent British English collected by the Survey of English Usage. (The set includes newspaper reports, student essays, novels and some essential information.) The figures below show how often certain words appear in the set.

Official terms	*Plainer alternatives*
accordingly *6*	so *281*
advices *0*	information *179*, instructions *18*
attain *3*	reach *113*
category *12*	group *130*
cease [and four tenses] *23*	stop [and same four tenses] *92*
commence *9*	begin *168*
concept *27*	idea *72*
entitlement *17*	right *59*
forward (verb) *0*	send *146*
initiate *15*	begin *168*
in the event of *7*	if *1288*
notwithstanding *3*	despite *55*
particulars *0*	details *54*
persons *20*	people *462*
prior to *13*	before *186*
regarding *24*	about *389*
terminate *11*	end *51*
utilize *3*	use *98*
whilst *57*	while *234*

The figures compare like with like, as far as possible. For example, 'accordingly' belongs to the grammatical class called co-ordinators, and the table gives figures for the number of times its plainer equivalent, 'so', is used as a co-ordinator.

A word on some troublesome words

address 'We are addressing the problem and will let you have a reply soon.' Occasionally ambiguity arises from the use of 'address' in this way: 'The minister has been addressing the letter he received from Mr Smith.' Cynics say the verb is often used as a cover for inaction. More precise verbs are usually available, such as 'consider' and 'tackle'.

advices Still a favourite pomposity among lawyers: 'I await your advices in respect of this matter.' Better to be precise and use 'instructions', 'suggestions' or 'comments'.

advise Near-universal business English for 'inform', 'tell' or 'let me know', but to avoid ambiguity it is better to use one of these and keep 'advise' for 'give advice to'. This prevents such officialese as 'please be advised that your cheque is in the post' which, apart from usually being untrue, begins with four redundant words.

and/or The oblique puts the reader to the trouble of creating three possible meanings. Instead of writing 'We will bring horses and/or donkeys', prefer 'We will bring horses or donkeys or both.'

anticipate Often used as a posh word for 'expect'. Their different meanings should not be allowed to merge, say careful users. 'Anticipate' means to take some action to forestall, or benefit from, a future event:

Customers anticipated the rise in interest rates by choosing fixed-rate loans.

apropos or **à propos** Prefer 'about', 'concerning' or 'with reference to'.

as and when This never seems to mean anything more than 'when' or 'if':

As and when the go-ahead is given, the project will involve the construction of a new underground station as well as rail tunnels.

This probably means:

As soon as we get the go-ahead for the project, we will begin constructing a new underground station and rail tunnels.

as at 'The value of your shares as at 5 April was £324.' Normally 'on' will do.

as per 'I enclose the form as per our conversation.' It's better, if not always possible, to avoid mixing English and Latin in the same phrase. Prefer 'I enclose the form as discussed.' Alternatively, 'in accordance with' will often do the job. 'As per usual' is especially cringe-making.

at the end of the day Waffle for 'ultimately', 'eventually' or 'in the end'. The only thing that can truly be said to occur at the end of the day is nightfall.

avail yourself 'If you fail to avail yourself of the facilities above, your electricity supply could be cut off without further notice.' Avoid such woolly, pseudo-genteel expressions. Prefer 'If you do not use [or 'take up' or 'make use of'] one of the repayment methods shown above . . .'.

can Careful users are very careful with 'can' and 'may'. The first means 'is able to' as in 'Helen can take solid food now that her jaw is mended.' The second means 'is permitted to', as in 'Helen may take solid food because the doctor says it's safe to do so.' The second meaning is common in plain (and not-so-plain) legal documents:

> **The secretary of state may make rules requiring a patents register to be kept. The rules may delegate the keeping of the register to another person.**

'May' can also imply a positive possibility:

> **You should show this letter to other joint shareholders as they may receive different information from elsewhere**

whereas 'might' in that position would suggest that the possibility is more remote. In practice most people use 'may' and 'might' interchangeably.

disconnect 'We regret that your supply could be disconnected unless you make a satisfactory arrangement to pay the arrears.' It used to be impossible to persuade electricity and gas suppliers to use the plainer 'cut off' – it was thought to be not quite nice. Now they all do it.

documentation 'We will let you have the documentation in the next ten days.' This is pompous: 'documents' will do.

duly If this means anything, it means 'in the correct manner'. A company's share-offer letter includes a rash of dulys:

> **I am pleased to tell you that . . . as a qualifying shareholder, an application for shares duly made by you . . . will receive special treatment . . . This means that, if you duly apply in the offer . . . If you duly apply on this application form . . .'**

The company could have avoided the word by saying on the application form that only correctly completed forms would qualify for the offer.

etc (Latin: et cetera, meaning 'the rest'.) This is perhaps the best understood Latin abbreviation and, though sometimes labelled '*e*xtreme *t*hought *c*ollapse', is harmless if used in moderation and if precision is not required:

The burglar stole televisions, videos etc.

If a list is introduced by a word like 'such as' or 'includes', 'etc' is unnecessary because the reader knows the list is incomplete:

The stolen goods included televisions, videos and stereos.

general consensus of opinion 'Consensus' covers it, in a word. And it needs three *s*'s.

get, got Primary-school teachers' exasperation at the overuse of these words has left many people afraid to use them at all in formal writing. The alternatives 'receive' and 'obtain' are plain enough but there's no harm in 'get' and 'got' (words with a thousand years of common use behind them) when you feel that their simplicity and informality are helpful:

I got your letter yesterday. Thank you for keeping me informed. I'll send someone around to fix the problem.

If your managers dislike the words, quote Proverbs at them:

Wisdom is the principal thing therefore get wisdom and with all thy getting get understanding

which doubles as a counterblast to people who say you shouldn't repeat the same word in a sentence or paragraph.

having said that A waffling way of writing 'but', 'however' or 'even so'. Avoid it. Addicts can withdraw slowly with 'that said'.

hence A useful but regrettably uncommon word. It can mean 'as a result' ('hence the election will take place next week') or 'from now' ('The election will take place two weeks hence'). Use the word regularly or it will disappear.

herewith, hereby, hereof, heretofore, hereinbefore, hereto, herein These smell of old law books; they are not plain words. 'I hereby declare' just means 'I declare'. 'The document attached hereto' just means 'the document attached' or 'the document attached to this'. Happily, these words are slowly disappearing from business and legal use. Such linguistic change often occurs and should cause no-one to fear the imminent death of the language. Do we miss, for example, a range of once-common words beginning with the 'wan-' prefix: 'wanthrift' (extravagance), 'wanhap' (misfortune), 'wantruth' (falsehood) and 'wanchance' (ill-luck)? Only 'wanton' (undisciplined) remains in fairly common use.

however As this is often used as a more leisurely alternative to 'but', it normally needs to be preceded and followed by a comma:

It seems clear that your opponent lied in court. We do not believe, however, that she is likely to be convicted of perjury.

If 'however' begins a sentence – often the best position, since it tells the reader immediately what sort of point is coming up – a comma will normally follow:

However, we do not believe that she is likely to be convicted of perjury.

This sentence shows a common mistake with 'however':

You have told us that the ring was stolen while the back door was left open, however, the policy only covers theft from your home if force is used to enter or leave.

Because 'however' seems to lengthen the pause that precedes it, a comma is not enough. A semicolon or full stop (see also chapter 11) would be better:

You have told us that the ring was stolen while the back door was left open. However, the policy only covers theft from your home if force is used to enter or leave.

Or you could use 'but' and save some of the punctuation:

You have told us that the ring was stolen while the back door was left open, but the policy only covers theft from your home if force was used to enter or leave.

Obviously there is no need for a comma in such phrases as 'However likeable he was, he had a vicious streak.'

I write or **I am writing** Many writers are told not to begin letters with these phrases because it's obvious they are writing. But 'I write to explain the department's policy on . . .' is quite harmless – you could not launch in with 'I explain . . .'. On the other hand 'I write to inform you that' might as well be deleted, as nothing useful has been said.

methodology This means a body of methods or the study of method and its application in a particular field. It is often misused as a posh word for 'method'.

Ms This, pronounced 'miz', is now the courtesy title of choice for many women who do not wish their marital status to be disclosed by 'Miss' or 'Mrs'. Use it unless the woman has asked to be addressed by a different title. If a woman signs her letters 'Susan Hopkins' without indicating a preferred title, there is no harm in writing to her as 'Dear Susan Hopkins' or 'Dear Ms Hopkins'.

null and void 'Null' means 'void' ('of no effect'), so there's no need for both words in a phrase like 'The agreement is null and void'. Simplest is 'void' or 'worthless'.

opine 'The barrister opined that the case would fail in court.' This unusual verb smells of pomposity in British English and is best avoided. This is a pity as it leads to wordiness like 'It is the opinion of the barrister that . . .'. Alternative verbs include 'believe', 'consider', 'say'.

prior to This pompous way of saying 'before' has escaped from its cage and bred plentifully. It tends to produce clumsy, verbless constructions: 'Prior to the abandonment of the mine by the company . . .' instead of 'Before the company abandoned the mine . . .'. It is better to keep 'prior' as an adjective – 'prior approval', 'prior discussion' – or as a noun for people in charge of monasteries.

re 'Re: Claim for housing benefit.' A Latin remnant from the word 'res' ('a thing'), this is not a short form of 'reference' or 'regarding'. The word should be struck out of all headings to letters – let them stand on their own. In text, prefer 'about', 'concerning' or 'on'. When referring to cases, lawyers often use 're' to mean 'in the matter of': 'Re [or even *In re*] Casaubon 1992'. Better is 'In Casaubon 1992'.

shall The old rule was that when writing of future events you would say 'I *shall*; you will; he/she/it will; we *shall*; you (plural) will; they will' but that when writing of promises, obligations or commands, the *wills* and *shalls* would change places. This is why the British coronation oath goes:

Archbishop: Will you to your power cause law and justice, with mercy, to be executed in all your judgments?

Sovereign: I will.

And it's why Binyon's poem of 1914, often quoted in memory of the war dead, says:

They shall grow not old, as we that are left grow old,
Age shall not weary them, nor the years condemn.
At the going down of the sun and in the morning
We will remember them.

But only one person in a million now understands these distinctions, so there is confusion when legal documents use 'shall' in an effort to impose an obligation. 'Must' is clearer for this purpose: 'The tenant must pay the rent on time.' Conveniently for people who take this view, the Old English root of 'shall' is 'sceal', meaning 'I must' or 'I owe'. The Survey of English Usage texts referred to earlier in this chapter show 90 uses of 'shall', so it is still common, but 362 of 'must'.

should Traditionally this was the correct form of the future conditional tense in the first person singular and plural. Or, in plain English, it was right to say 'I *should*; you would; he/she/it would; we *should*; you would; they would.' This is why many still prefer to write: 'I should be grateful if you would . . .' not 'I would be grateful if you could . . .' or 'I would be grateful if you would . . .'. A neat sidestep is to write 'I'd be grateful if you would . . .' or 'Would you please . . .'.

to hand 'I have to hand your letter of 15 January.' This is the kind of snooty, you-are-a-microbe language that gets bureaucrats a bad name. Avoid it with 'thank you' or 'I refer to'.

unless and until 'Unless and until the conditions are met, the deal is off.' This usage smells of officialdom. Either 'unless' or 'until' will do.

whilst This is becoming unusual (especially in American English). Its job has been usurped by 'while', with which it is interchangeable in standard English.

Foreign words

English is certainly the richer for its contact with foreign words, with hundreds of direct borrowings like 'caravan', 'trek', 'graffiti', 'bastard', 'clan', 'crag', 'criterion', 'phenomenon', 'slogan', 'corgi', 'garage' and 'armada'. Then there are thousands of foreign-derived words such as 'plant', 'fantasy', 'custom', 'interest', 'jury', 'mutton' and 'tea'. (For an account of the history of such words, see any large single-volume dictionary.)

It's not always easy to judge what will be clear to readers of essential information. 'Ad hoc', in such phrases as 'ad hoc group' and 'ad hoc committee', fills a useful gap but could hardly be considered plain to most people. The terms 'ie' and 'eg' will be understood by a professional audience and benefit from being concise, but for a mass audience the English equivalents are safer. Using uncommon foreign-language terms may look like showing off, so unless you are sure of your audience avoid those words so often seen in the literary review pages like 'oeuvre', 'Bildungsroman' and 'auteur'. If you want your staff to provide a guide or a briefing note, don't ask them for a 'vade mecum'.

Some terms haven't entirely lost their strangeness but have made the transition into plain English. In this group are 'vice versa', which saves the writer many words of explanation (though 'the other way round' is sometimes good enough), 'per cent' and 'etc'. 'Curriculum vitae' seems likely to endure, shortened to 'CV'; the best alternative that American English can come up with is French – résumé.

Latin still fights a rearguard action on the British pound coin, where it is more heavily represented than English or Welsh with phrases like 'Nemo me impune lacessit' ('No-one harms me and gets away with it') and 'Decus et tutamen' ('Glory and protection'). Latin's conciseness undoubtedly helps when space is tight, but its use on the currency suggests that the plain languages of the UK are not thought good enough for some ceremonial purposes.

Latin or French [F] phrase	*Meaning or alternative phrase*
ad hoc	for this purpose or occasion
carte blanche [F]	a free hand, freedom
ceteris paribus	other things being equal
cf (conferre)	compare
circa	about
de minimis	trivialities, small amounts
eg (exempli gratia)	such as, for example
en bloc [F]	as a whole, together
etc (et cetera)	and so on, and the rest
ex officio	by virtue of the office held
ibid (ibidem)	in the same place, book etc
ie (id est)	that is
inter alia/alios	among other things/people
modus operandi	way of working, method
mutatis mutandis	with the necessary changes
op cit (opus citatum)	work quoted
per annum	per year, a year, annually
per capita	per head, per person, each
per diem	per day, a day, daily
per se	as such, by or in itself, essentially
pp (per procurationem)	on behalf of, by the agency of
pro forma	a form
qv (quod vide)	see
seriatim	one at a time; in the same order
sic	thus! (drawing notice to error)
sine die	indefinitely
vis-à-vis [F]	as regards, regarding, on, about
viz (videlicet)	namely
vs (versus)	v, versus, against
vs (vide supra)	see above

3 Writing tight

Guideline: *Use only as many words as you really need.*

Flab. Writing is full of it. These 95 words from a loan company tell borrowers that they have fallen behind with their repayments. More than one third can be cut without losing any meaning:

> **Arrears at present subsist on your mortgage account in the sum of £1,032, with a further payment becoming due on the 11th April. In view of the account being a mortgage account, we are not in a position to stop interest being debited each month and in order to prevent the account situation from deteriorating, it is necessary that payments are received each month which represent the interest debit. At present this amount is £242 and therefore it is regretted your offer to make payments in the sum of £80 a month is not sufficient.**

You may like to blue-pencil it yourself, but at the end of the chapter we'll see how so many words can be saved.

Part of writing well is writing tight, ruthlessly cutting dross. Most readers are busy people who want to find out the main points of your message, and fast. Making them read excess words is an unfriendly act, especially in business where a deluge of unwanted paper falls on everyone at every level.

Not that shorter is always better; sometimes you need more words to make a point clear. Plain, certainly. Tight, certainly. But not so plain or tight that you miss out essential points or come across as blunt and rude. Being ruthless with words needn't mean being graceless with people.

Cutting dross allows your information to shine more clearly. In the early 1900s, Professor William Strunk used to tell his students: 'Omit needless words, omit needless words, omit needless words.' (Once should have been enough, but he was keen.) He believed that just as drawings should have no unnecessary lines and machines no unnecessary parts, so sentences should have no unnecessary words.

Easy to agree with, perhaps, but hard to do. The key is to let the first draft stand as long as possible, then return and revise it. Then revise it again. And probably again. In business, of course, time is against you: that letter or report must go out tonight. And useless words aren't always obvious – they have to be hunted. So let's examine the three main techniques for dealing with them:

1. Striking out useless words.
2. Pruning the dead wood, grafting on the vigorous.
3. Rewriting completely.

Striking out useless words

The most obviously useless words are straight repetition:

The cheque that was received from Classic Assurance was received on 13 January.

'Was received' occurs twice, so the sentence could say

The cheque ~~that was received~~ from Classic Assurance was received on 13 January

or

The cheque ~~that was received~~ from Classic Assurance ~~was received~~ arrived on 13 January.

Spotting this kind of problem becomes harder as the distance between repetitions increases:

The standard of traffic management on the A57, A59 and A623 is of a lower standard than on other major roads in the region.

It doesn't make sense to say that a 'standard . . . is of a lower standard', so the rewrite would be:

The standard of traffic management on the A57, A59 and A623 is ~~of a~~ lower ~~standard~~ than on other major roads in the region.

Then there are words that repeat an idea:

We attach herewith a financial statement.

If the statement is attached – or enclosed – it must be 'herewith', so the word can disappear:

We attach ~~herewith~~ a financial statement.

Wordiness often comes from trying to make a simple procedure sound impressive:

A new bank account is in the process of being set up for you.

Delete four words and this becomes:

A new bank account is ~~in the process of~~ being set up for you.

The verb 'carry out' (like 'undertake' and 'perform') always merits suspicion; often a more vigorous expression will make the same point more economically. For example, it can simply be deleted:

Work is required to be carried out on the flue and funnels

becomes

Work is required t~~o be carried out~~ on the flue and funnels.

Or 'carry out' can be cut by strengthening the verb it supports:

The firm does not intend to remove the lime trees but it is necessary to carry out pruning to the trees to keep them healthy.

This becomes:

The firm does not intend to remove the lime trees but it is necessary to ~~carry out pruning to~~ prune the trees to keep them healthy

or

The firm does not intend to remove the lime trees but they need to be pruned to keep them healthy.

It's not always so easy to see redundancy. Take this sentence:

For the benefit of new members, the secretary described the rules of the committee and the remit that had been given to it.

Since 'remit' means the committee's terms of reference and a remit must, by its nature, be 'given', the last six words are redundant:

For the benefit of new members, the secretary described the rules of the committee and ~~the~~ its remit ~~that had been given to it~~.

Cut out useless phrases like 'it should be pointed out that', 'it must be noted that', 'I should mention that', 'I would inform you that' and 'I would stress that'. Better just to point it out, note it, mention it, or stress it. In the following examples, the useless words are underlined; what remains makes perfectly good sense on its own:

I must point out that I am legally obliged to charge rates on the property's current value.

I would like to take this opportunity to apologize for the delay in replying to your complaint.

It should be appreciated that there is always an element of under-reporting of accidents, particularly if no-one is injured. It should also be noted that our accident figures exclude occurrences where the system for explosion relief operated effectively.

It is only fair for me to point out at this point that the committee showed great concern about your case at its last meeting.

This letter is to advise you that, following our successful seminar last year for suppliers, we plan to hold another on 13 October.

Pruning the dead wood, grafting on the vigorous

Dead wood often makes a fine pretence of being alive, so a keen eye is needed. In this example, ten unnecessary words could be replaced by one:

May I draw your attention to the final account dated 28 June from which I note that six payments of £18 were credited to your account from 28 March to 25 August, totalling £108.

The first six words are courteous but they delay the main message unduly and could go. Then, 'from which I note' is pompous – what matters is not what the writer notes but what the final account shows. Using the vigorous verb 'shows', the sentence becomes:

~~May I draw your attention to~~ the final account dated 28 June ~~from which I note~~ shows that six payments of £18 were credited to your account from 28 March to 25 August, totalling £108.

Certain words are prime suspects for pruning. They include 'situation', 'aspect', 'facility', 'issue', 'element', 'factor', 'matter' and 'concept'. All are occasionally useful but they tend to be overworked at the expense of more concrete words:

The company has one engineer on call at all times, giving excellent speed of response in emergency situations.

An 'emergency situation' is nothing more than an emergency, so 'situation' can safely be cut:

The company has one engineer on call at all times, giving excellent speed of response in emergencies.

Another phrase often accompanied by verbiage is 'the fact'. There is 'given the fact that' and 'in the light of the fact that' (which just mean 'as' or 'since'); 'despite the fact that' (which means 'although'); and 'the fact of the matter is' (which, if it means anything, means 'the fact is'). Here, the six cancelled words could simply be replaced by 'as':

~~In view of the fact that~~ the central heating was fitted by Union Gas, they have cancelled the bill in the interests of good customer relations.

Rewriting completely

When there are far too many words for the message but neither of the first two methods will work, a total rewrite is the only alternative. Various signals may alert you to this need:

- The meaning isn't clear.
- The sentence is long and the verbs are few.
- The verbs are feeble – for example, they are smothered by nouns (see chapter 5), they are in the passive voice (see chapter 4), or they are derived from 'to be' or 'to have' (see chapter 4).

For example, an engineer is writing about the cost of materials for a road scheme:

Over-estimating on one type of material could have a detrimental cost effect for the clients, depending on the prices in the Bill of Quantities.

Alarm bells ring at 'have a detrimental cost effect'. First because 'have' seems feeble as the solitary verb in the sentence, and second because 'detrimental cost effect' is a pompous way of expressing the simple idea that the clients might have to pay more. So the whole sentence could say:

Over-estimating on one type of material could cost the clients more, depending on the prices in the Bill of Quantities.

While this saves only four words, there is now a strong verb, 'cost', and the message is immediately apparent.

An insurance firm is thinking of publishing a guidance booklet for managers of company car fleets. Its internal report on the idea begins:

AIM OF PROPOSED CAR FLEET MANAGEMENT GUIDE

This guide would have the objective of highlighting to car fleet managers the best way to achieve, and the benefits of adopting, a professional approach vis-à-vis managing a car fleet.

There are only a few publications at present covering the subject of car fleet management and with no current insurance company involvement there would appear to be a definite market niche for us to explore.

This makes sense but is far too flabby:

- 'Would have the objective of highlighting' is a wordy way of saying 'aims' or 'seeks'.
- In plain words, 'vis-à-vis' means 'towards' or 'to'.
- Sentences beginning 'there are', 'there is' and 'there were' are often wordy and reduce the strength of any remaining verbs. Two 'there' verbs are present here.
- In 'the subject of car fleet management', the first three words are redundant because readers know that 'car fleet management' is a subject.
- 'No current insurance company involvement' smothers the verb 'involve' (see chapter 5). In any case, a more expressive verb would be 'publish' or 'produce'.
- 'Insurance company' could be cut to 'insurer'.

So a first redraft might say:

AIM OF PROPOSED CAR FLEET MANAGEMENT GUIDE

This guide would show car fleet managers how they could best achieve a professional approach to managing a car fleet and the benefits of doing so.

At present, only a few publications cover car fleet management. None of them are produced by insurers so there is a definite market niche for us to explore.

A second redraft would go a little further:

AIM OF PROPOSED CAR FLEET MANAGEMENT GUIDE

This guide would show car fleet managers how to do their work more professionally and why this would benefit them.

Few publications cover car fleet management, none of them produced by insurers. So there is a definite market niche for us to explore.

The original had 72 words, this has 52. What remains is tight and doesn't waste the reader's time.

Putting it all together

The chapter began by promising to cut more than a third from a letter to a mortgage payer. First, useless words can go:

> **Arrears ~~at present~~ subsist on your mortgage account ~~in the sum~~ of £1,032, with a further payment ~~becoming~~ due on ~~the~~ 11~~th~~ April. ~~In view of the account being a mortgage account,~~ we are not in a position to stop interest being debited ~~each month~~ and ~~in order~~ to prevent the account ~~situation~~ from deteriorating, it is necessary that payments are received each month which represent the interest debit. At present this amount is £242 ~~and~~ therefore it is regretted that your offer to ~~make~~ pay~~ments in the sum of~~ £80 a month is not sufficient.**

Then vigorous words (underlined) can be grafted on:

> **The arrears ~~subsist~~ on your mortgage account ~~of~~ are £1,032, and ~~with~~ a further payment is due on 11 April. Regrettably we ~~are not in a position to~~ cannot stop interest being ~~debited~~ charged. Therefore, to prevent the ~~account from deteriorating,~~ arrears growing, ~~it is necessary that~~ you will need to pay the interest charge ~~are received~~ each month ~~which represent the interest debit~~. At present this ~~amount~~ is £242, ~~therefore~~ so ~~it is regretted~~ we regret that your offer to pay £80 a month is not sufficient.**

So the final version is:

> **The arrears on your mortgage account are £1,032, and a further payment is due on 11 April. Regrettably we cannot stop interest being charged. Therefore, to prevent the arrears growing, you will need to pay the interest charge each month. At present this is £242, so we regret that your offer to pay £80 a month is not sufficient.**

This is only 59 words, a cut of 37 per cent. It delivers the same facts and is just as courteous – perhaps more so.

Results from the focus group showed a strong preference for the final version over the original. The final version scored an average clarity mark of 17 points out of a possible 20, as against only 10 for the original. Twenty-nine people out of 34 preferred the final version.

4 Favouring the active voice

Guideline: *Prefer the active voice unless there's a good reason for using the passive.*

This sentence has an active voice verb:

Fred is demolishing the building

while this has a passive voice verb:

The building is being demolished by Fred.

This chapter explains:

- the difference between active and passive;
- how to convert one to the other;
- why the active should be your first choice;
- how 'I' and 'we' can make formal reports more readable;
- when the passive can be useful.

During a series of plain language workshops in India, I was surprised that so many of the lawyers and government officials who attended could readily distinguish between active and passive voice verbs. Most of them had learned English as a second language and had had to master the basics of its grammar. Yet in countries where English is the first language, 'active' and 'passive' are often alien terms that native English speakers only meet when they study a second language. This is a pity, as the ability to recognize and use the active voice helps to foster a plain English style.

Many writers have damned the passive voice unreservedly. In 1946 George Orwell wrote: 'Never use the passive where you can use the active', but this is going much too far. Certainly the active tends to make the writing tighter, more personal, and introduces action earlier in sentences. The passive tends to do the reverse yet is still a valuable tool, as we will see.

The words 'passive' and 'active' are well understood in their everyday meanings: 'Some men take an active role in infant care, but many are passive.' These everyday meanings differ from their grammatical ones. So, as you read this chapter, try to forget about the everyday meanings.

Recognizing active voice verbs ('active verbs', for short)

Putting the 'doer' – the person or thing doing the action in the sentence – in front of its verb will usually ensure that the verb is in the active. The following sentences all have active verbs (underlined) because the doers precede the verbs they govern:

The President <u>wants</u> an improved health service.

I <u>walked</u> up the stairs.

Armies <u>are</u> on the march.

She <u>hates</u> going to work.

Ice-cream <u>tastes</u> revolting.

Most of us favour active verbs when we speak. People would think you were odd if you continually said things like 'The house is being bought by me' (passive) instead of 'I am buying the house' (active), though the meaning is the same if they are spoken with the same stresses.

Occasionally a verb can be followed by a doer yet remain active:

She used to hate going to work, said her sister.

Here 'her sister' follows 'said' but the verb is still in the active because 'said her sister' is grammatically identical to 'her sister said'.

Recognizing passive voice verbs ('passive verbs', for short)

In most sentences with a passive verb, the doer follows the verb or is not stated, as here (verbs underlined):

(a) Three mistakes <u>were admitted</u> by the director.

(b) Coastal towns <u>are damaged</u> by storms.

(c) Verdicts <u>will</u> soon <u>be delivered</u> in the Smith case.

In (a) and (b), the doers ('director' and 'storms') follow the verbs through which they act. In (c) the doer is not stated; no-one can tell who or what will give the verdicts.

To put (a) and (b) into the active, bring the doer to the start of the sentence:

The director admitted three mistakes.

Storms damage [are damaging] coastal towns.

To convert (c) into the active, you would need to know the doer:

[The judge] will soon deliver verdicts in the Smith case.

An almost infallible test for passives is to check whether the verb consists of:

- part of the verb 'to be' (though this is sometimes implied rather than stated, as in many newspaper headlines where space is tight – 'Explorer [is] attacked in jungle'); and
- a past participle.

This is an easy test to apply if you can already recognize parts of 'to be' and past participles. If not, here's how:

Parts of the verb 'to be'

Present tense (explained in chapter 13): am, is, are, am being, is being, are being.
Past tenses: was, were, has been, have been, had been.
Future tense: will be, shall be.
Infinitive (explained in chapter 13): to be.

Be careful, however, not to confuse parts of 'to be' with parts of 'to have', which are 'has', 'had' and 'have'. The difference is clear from the expressions 'You are a frog' and 'You have a frog'. The first is about being, the second possession.

Past participle

All verbs have a past participle. To find it, begin with the infinitive form of the verb, for example:

to attract **to annoy** **to decide** **to go**

Cross out 'to' and send the rest into the past tense:

~~to~~ attract+ed   ~~to~~ annoy+ed   ~~to~~ decide+d   ~~to~~ go+ne or went

The first three are past participles. In the fourth, you need to apply a tie-break by putting 'we have' in front of each phrase, producing 'we have gone' and 'we have went'. Only the first of these makes sense in standard English, so 'gone' is the past participle while 'went' is just a past tense. The tie-break is useful because some common verbs have two candidates for past participle: 'see' (seen/saw); 'eat' (eaten/ate); 'take' (taken/took); and 'give' (given/gave). Applying the tie-break reveals that the first alternative is the past participle in each case.

Applying the full test for passives to the example sentences, it is clear that they all fulfil its criteria:

(a) Three mistakes were admitted by the director.

(b) Coastal towns are damaged by storms.

(c) Verdicts will soon be delivered in the Smith case.

In (c) the fact that 'soon' is splitting the verb makes no difference – the verb is still passive.

Computerized grammar checkers search for passives by applying the same test. Unfortunately this will occasionally throw up phantom passives. For example, 'is tired' would be flagged as a passive in the sentence:

A man who is tired of London is tired of life

as it appears to fulfil the criteria of having part of 'to be' and the past participle of 'to tire'. But it cannot be a passive because no doer can be attached to 'is tired' and none is implied. 'Tired' in each case is a description of the man, not part of the verb. The verb 'is' is in the active voice.

Converting passives to actives

Though using the active in the examples that follow will produce only small gains in clarity and economy, a general preference for active over passive will significantly improve the readability of most documents.

A financial adviser writes to his client:

We have been asked by your home insurers to obtain your written confirmation that all their requirements have been completed by yourself.

By applying the test, 'have been asked' and 'have been completed' are revealed as passives. Use of the active, putting the doers in front of the verbs, would give:

Your home insurers have asked us to obtain your written confirmation that you have completed all their requirements.

I wanted to see whether the focus group would record any preference between these two versions, as the only difference is in the use of active and passive. Though both were regarded as clear (13 points out of 20 for the passive sentence, 17 for the active), 28 people out of 35 preferred the active.

A safety official writes (passives underlined):

A recommendation <u>was made</u> by inspectors that consideration <u>be given</u> by the company to the fitting of an interlock trip between the ventilation systems to prevent cell pressurisation.

Converting passive to active, the sentence becomes (actives underlined):

Inspectors <u>made</u> a recommendation that the company <u>give</u> consideration to the fitting of an interlock trip between the ventilation systems to prevent cell pressurisation.

Then, using the strong verbs hidden beneath 'recommendation' and 'consideration', the sentence becomes even crisper – and ten words shorter than the original:

Inspectors <u>recommended</u> that the company <u>consider</u> fitting an interlock trip between the ventilation systems to prevent cell pressurisation.

In the focus group, 18 people out of 35 preferred this final version over the passive sentence. Nine preferred the passive.

Verbs provide so much useful information that readers prefer to get them early in sentences; this tends not to happen when the verbs are passive. Placing an important verb late forces readers to store large chunks of text in their short-term memory while they wait to discover the doer and what the action will be. The problem is worsened if there are other hurdles, like brackets containing exceptions and qualifications:

If you decide to cancel your application, a cheque for the amount of your investment (subject to a deduction of the amount (if any) by

which the value of your investment has fallen at the date at which your cancellation form is received by us) will be sent to you.

'You decide' and 'has fallen' are active, while 'is received' and 'will be sent' are passive. 'Will be sent' needs converting into the active because it is too far (36 words) from the noun it refers to, which is 'cheque'. Making this single change would produce:

If you decide to cancel your application, <u>we will send you</u> a cheque for the amount of your investment (subject to a deduction of the amount (if any) by which the value of your investment has fallen at the date at which your cancellation form is received by us) ~~will be sent to you~~.

This would be the first stage in a comprehensive kill of brackets and other debris that would produce:

If you decide to cancel your application, we will send you a cheque for the amount of your investment less any fall in its value at the date we receive your cancellation form.

This is a third shorter than the original. Using the active voice has enabled other healthy writing habits to come into play.

Using 'I' or 'we' in formal reports

The myth that 'I' and 'we' should be avoided in formal reports has crippled many writers, causing them to adopt clumsy and confusing constructions like referring to themselves as 'the writer' or using impersonal passives like 'it is thought', 'it is felt', 'it is believed', 'it is understood' and

It is considered that fluoridation of drinking water is beneficial to health

from which readers have to guess who is expressing the view: the writer, wider scientific opinion, public opinion, or all three. In reports, readers should not have to guess. Attempts to prohibit 'I' and 'we' are particularly strange in that any other person, creature or thing may be mentioned in a report.

If you are writing on your own behalf, use 'I' and 'my' judiciously, but don't overdo it for fear of seeming big-headed. If you are writing on behalf of two or more people, let 'we' and 'our' do the same job. (If there is some overriding reason why these tactics are impossible, you should still make sure that most of your sentences have doers – perhaps the name of your section or organization.)

Writers of scientific and technical material will especially benefit from using 'I' and 'we', which are becoming commonplace in many journals including the British Medical Journal. Unless a journal specifically prohibits these words – and most do not – you should feel free to use them. Almost everyone who writes about scientific and technical writing recommends this. Don't be seduced by the idea that impersonal writing makes you sound more scientific: no-one ever became a scientist by wearing a white laboratory coat.

Warning: passives can be useful

Passive verbs have their uses and it would be silly – as well as futile – for the style police to outlaw them. There are five main reasons for using them:

- To defuse hostility – actives can sometimes be too direct and blunt.
- To avoid having to say who did the action, perhaps because the doer is irrelevant or obvious from the context.
- To focus attention on the receiver of the action by putting it first – 'An 18-year-old girl has been arrested by police in connection with the Blankshire murders.'
- To spread or evade responsibility by omitting the doer, for example 'Regrettably, your file has been lost.'
- To help in positioning old or known information at the start of a sentence or clause, and new information at the end.

The last point relates to an important benefit of the passive. Read these two sentences about a nuclear reactor:

Concern has been raised about arrangements for gaining immediate access to the chimney. Winch failure or the presence of debris between the platform edge and the chimney internal wall may necessitate access.

The second sentence, written in the active voice, doesn't seem to follow from the first, whose primary focus – placed late in the sentence – is about gaining access. Now try it with the second sentence in the passive voice:

Concern has been raised about arrangements for gaining immediate access to the chimney. Access may be needed if the winch fails or there are debris between the platform edge and the chimney internal wall.

The topic of the second sentence, access, is now introduced early in that sentence and developed. There is a clear link between the focus of the first sentence and the topic of the second – a common device that helps the writing to flow by taking the reader from the known to the unknown.

Checking your passive percentage

You can check how many passives you use with a computerized grammar checker, or manually. Your passive percentage is given by the formula:

100 (Number of passives ÷ number of sentences)

If you score over 50 per cent (that's one passive every two sentences) check your verbs carefully. Do you really need so many passives?

5 Using vigorous verbs

Guideline: *Use the clearest, crispest, liveliest verb to express your thoughts.*

Good verbs give your writing its power and passion and delicacy. It is a simple truth that in most sentences you should express action through verbs, just as you do when you speak. Yet in so many sentences the verbs are smothered, all their vitality trapped beneath heavy noun phrases based on the verbs themselves. This chapter is about releasing the power in these smothered verbs.

Business and official writing uses plenty of smothered verbs:

- People don't *apply* for a travel pass, they make application.
- Speakers don't *inform* the public, they give information.
- Officials don't urgently *consider* a request, they give it urgent consideration.
- Staff don't *evaluate* a project, they perform an evaluation.
- Scientists don't *analyse* data, they conduct an analysis.
- Citizens don't *renew* their library books, they carry out a process of library book renewal.

In each case the simple verb (in italics) is being converted into a noun that needs support from another verb. The technical term for a noun that masks a verb in this way is *nominalization*. There is nothing wrong with nominalization as such – it is a useful part of the language. But overusing it tends to freeze-frame the action.

The following examples show how vigorous verbs can improve sentences containing nominalizations, making them more powerful and concise. Three types of construction are considered:

- Nominalization linked to parts of 'to be' or 'to have';
- Nominalization linked to active verbs or infinitives;
- Nominalization linked to passive verbs.

Nominalization linked to parts of 'to be' or 'to have'

Parts of 'to be' include 'are', 'is', 'was', 'were', 'has been' and 'have been'. Parts of 'to have' include 'has', 'had' and 'have'.

Many sentences whose only verbs are parts of 'to be' or 'to have' are perfectly clear and crisp, for example:

All animals are equal, but some animals are more equal than others.

Detailed management information is available.

In other sentences, the linking of such a verb with a nominalization is a good reason for suspecting that improvement is possible, as in this example from an official letter:

I have now had sight of your letter to Mr Jones.

The main verb is 'had' while the nominalization is 'sight', which smothers 'see'. So it would be simpler to say:

I have now <u>seen</u> your letter to Mr Jones.

This example is from a business letter:

Funding and waste management have a direct effect on progress towards the decommissioning of plant and equipment.

The main verb is 'have' while the nominalization is 'effect', which smothers 'affect'. It is crisper to write:

Funding and waste management <u>directly affect</u> progress towards the decommissioning of plant and equipment.

A combination of part of 'to be' and a nominalization is easy to see in this example:

The original intention of the researchers was to discover the state of the equipment.

The nominalization is 'intention', smothering 'intend', while 'was' acts as a prop. A revision would say:

Originally the researchers intended to discover the state of the equipment.

Nominalization linked to active verbs or infinitives

This construction is easy to rewrite as the presence of active verbs usually means that the word order can be preserved.

A group of conservationists is writing to a local government department:

The group considers that the director of community services should proceed with the introduction of as many mini-recycling centres as the budget allows.

The nominalization is 'introduction', propped up by 'proceed'. The rewrite would use the active voice to revive the smothered verb 'introduce':

The group considers that the director of community services should introduce as many mini-recycling centres as the budget allows.

A company report explains what some of the staff do:

The team's role is to perform problem definition and resolution.

Two nominalizations, 'definition' and 'resolution' are propped up by 'to perform'. Using the smothered verbs 'define' and 'resolve', this becomes:

The team's role is to define problems and resolve them.

or

The team's role is to define and resolve problems.

A government department writes:

The policy branch has carried out a review of our arrangements in order to effect improvements in the reporting of accidents.

Here there are two nominalizations, 'review' and 'improvements', supported by 'carried out' and 'effect' respectively. Using vigorous verbs would produce:

The policy branch has reviewed our arrangements in order to improve the reporting of accidents.

Nominalizations linked to passive verbs

This construction is harder to revise because changing from passive to active disrupts the original word order. To compensate, however, the satisfaction is usually greater.

A housing association writes:

Notification has been received from the insurers that they wish to re-issue the Tenants Scheme Policy.

The nominalization is 'notification' and the passive is 'has been received'. Using the smothered verb 'notify' produces:

The insurers have notified us that they wish to re-issue the Tenants Scheme Policy.

A safety officer writes:

An examination of the maintenance records for the plant was carried out by Mr Patel.

This becomes, by the same technique:

Mr Patel examined the maintenance records for the plant.

When the going gets tougher

Sometimes the difficulties are harder to spot. But remember the common signals – nominalizations (often ending in *-ion*), passive voice, the verbs 'make' and 'carry out', and verbs derived from 'to be' or 'to have'.

A local government department writes:

The committee made a resolution that a study be carried out by officials into the feasibility of the provision of bottle banks in the area.

The rewriting task can oe split into three operations:

1 The nominalization 'resolution' becomes a strong verb: 'The committee *resolved* that'.

2 The nominalization 'study' becomes the active verb 'study': 'officials should *study*'.

3 The nominalization 'provision' becomes a present participle (see chapter 13): 'the feasibility of *providing* bottle banks in the area'.

So the complete result reads:

The committee resolved that officials should study the feasibility of providing bottle banks in the area

or

The committee resolved that officials should investigate whether providing bottle banks is feasible in the area.

An accountant writes:

The incidence of serious monetary losses in several transactions entered into by the firm during the year is causing us great concern.

'Incidence', meaning 'rate of occurrence', isn't derived directly from a verb. Here it is probably used as a posh way of saying 'occurrence'. The real verb should be 'occur', so a rewrite could say:

The serious monetary losses that have occurred in several transactions entered into by the firm during the year are causing us great concern.

Then, putting the verb at the start of the sentence and tidying up the rest, the result would be:

We are greatly concerned about the serious monetary losses that have occurred in several of the firm's transactions during the year.

This is 21 words long, compared with 22 in the original. This would be too small a gain to justify the effort, if brevity was the only criterion. Much more important is that the sentence can now be read without stumbling.

6 | Using vertical lists

Guideline: *Use vertical lists to break up complicated text.*

Vertical lists have become a common feature of many documents since the 1970s, helping to present complex information in manageable chunks. For example, instead of saying this:

> **Our inspections will be targeted on food factories. Inspectors will investigate food factory performance, establish each occupier's performance, establish the occupier's knowledge of health and safety risks in the industry and identify which other parts of our organization could provide us with support for law enforcement**

we could say this:

> **Our inspections will be targeted on food factories. Inspectors will:**
>
> - **investigate food factory performance;**
> - **establish each occupier's performance;**
> - **establish the occupier's knowledge of health and safety risks in the industry; and**
> - **identify which other parts of our organization could provide us with support for law enforcement.**

Though this takes up more space it is very easy to grasp.

Vertical lists can cause problems in three areas:

- keeping the listed items in parallel;
- punctuating the listed items;
- numbering the listed items.

Let's examine each in turn.

Keeping the listed items in parallel

A dietitian is explaining how a patient should cut her salt consumption:

To restrict your salt intake, you should:

- **not add salt at the table;**
- **use only a little salt in cooking;**
- **do not use bicarbonate of soda or baking powder in cooking;**
- **avoid salty food like tinned fish, roasted peanuts, olives.**

All the listed items are understandable individually and they are all commands, so to that extent they are in parallel. But the third point doesn't fit with the lead-in or 'platform' statement. Together they are saying:

you <u>should do not use</u> bicarbonate of soda or baking powder in cooking

which is nonsense. Obviously, 'do' should be struck out to create a true parallel structure. It might then be a good idea to group the positive and negative commands:

To restrict your salt intake, you should:

- **not add salt at the table;**
- **not use bicarbonate of soda or baking powder in cooking;**
- **use only a little salt in cooking;**
- **avoid salty food like tinned fish, roasted peanuts, olives.**

Unfortunately the job isn't complete, as there is now an odd mixture of positive and negative. The best solution might be to shift the remaining positive statement into the platform:

To restrict your salt intake, you should only use a little salt in cooking and you should not:

- **add salt at the table;**
- **use bicarbonate of soda or baking powder in cooking;**
- **eat salty food like tinned fish, roasted peanuts, olives.**

Sometimes a platform needs to be created to maintain the parallel structure. Here, a clerk is being told how to do a task:

- **You should check that the details on the self-certificate or medical certificate match those on the person's information card.**

- **You should check that the certificate has been completed correctly and conforms to the rules on validity.**
- **That the certificate covers the period of absence.**

By the third item the writer must have become tired of writing 'you should check that'. Rather than omit it, he or she should have converted it into a platform, producing:

You should check that:

- **the details on the self-certificate or medical certificate match those on the person's information card;**
- **the certificate has been completed correctly and conforms to the rules on validity;**
- **the certificate covers the period of absence.**

Often a vertical list is easier to read if each listed item has a similar grammatical structure. For example, they could all be statements that begin with infinitives or active verbs or passive verbs or present participles (see chapter 13 for explanations). In this list, all the listed items are passive verb statements, underlined for ease of reference:

The inspector should check that:

- **the vehicle is properly marked with hazard plates;**
- **the engine and cab heater are switched off during loading and unloading of explosives;**
- **any tobacco or cigarettes are kept in a suitable container and matches or cigarette lighters are not being kept in the cab;**
- **the explosives are securely stowed;**

There would be no harm in adding some active voice statements as long as they made sense when linked to the platform:

- **there are no unsecured metal objects in the vehicle's load-carrying compartments;**
- **the vehicle is carrying one or more efficient fire extinguishers.**

When statements with different grammatical structure are mixed haphazardly, the reader has to stop and backtrack. In this example, the listed items have infinitives, actives, passives or no verbs at all:

When the committee's work began, it established the following aims:

- **make the regulations simple to understand and up to date in structure and layout;**

- **to update forms and leaflets where necessary with details of current fees;**
- **the effects of competition will be considered;**
- **the creation of a document summarising details of the regulations which will enable people to focus on key issues and requirements;**
- **recent changes in legislation should be taken into account.**

As these points are supposed to be aims, they could all be put into the infinitive:

When the committee's work began, it established the following aims:

- **to make the regulations simple to understand and up to date in structure and layout;**
- **to update forms and leaflets where necessary with details of current fees;**
- **to consider the effects of competition ~~will be considered~~;**
- **to create ~~the creation of~~ a document summarising details of the regulations which will enable people to focus on key issues and requirements;**
- **to take account of recent changes in legislation ~~should be taken into account~~.**

Readers get used to the pattern here; they can then concentrate better on the meaning.

Punctuating the listed items

Vertical lists need punctuating as consistently as possible so that readers get used to a pattern and are not distracted by deviations. Here is a typical example of inconsistency:

The new job-holder will:

- **develop a set of guidelines for clean wastepaper recycling**
- **Introduce green bins for clean wastepaper at appropriate places;**
- **monitor compliance with departmental targets.**

Two of the listed items begin with a lower-case letter and one with a capital. One of them ends with a semicolon, another with a full stop, and the first with nothing at all.

For greater consistency, I suggest a two-part standard.

The first part of the standard is that when a listed item is a sentence or sentence fragment that relies on the platform statement to give it meaning, it should begin with a lower-case letter and end with a semicolon – except for the final item, which should normally end with a full stop. This produces the following result:

The new job-holder will:

- **develop a set of guidelines for clean wastepaper recycling;**
- **introduce green bins for clean wastepaper at appropriate places;**
- **monitor compliance with departmental targets.**

If you wish, you could add 'and' after each of the first two semicolons or, more conventionally, after the final semicolon only. If you wanted to show that only one of the jobs had to be done, you would put 'or' after the first two semicolons or, more conventionally, after the final semicolon only:

The new job-holder will:

- **develop a set of guidelines for clean wastepaper recycling;**
- **introduce green bins for clean wastepaper at appropriate places; or**
- **monitor compliance with departmental targets.**

Alternatively you could use this kind of set-up, which stresses the 'or':

The new job-holder will:

- **develop a set of guidelines for clean wastepaper recycling;**

or • **introduce green bins for clean wastepaper at appropriate places;**

or • **monitor compliance with departmental targets.**

Rarely will you want to continue a sentence beyond a list as this could overburden the reader's short-term memory. If you do, your last listed item should end with a semicolon and the sentence should continue with a lower-case letter:

The new job-holder will:

- **develop a set of guidelines for clean wastepaper recycling;**
- **introduce green bins for clean wastepaper at appropriate places; and**
- **monitor compliance with departmental targets;**

but the work must always take place within existing budgetary restrictions.

The second part of the standard applies to listed items that are complete sentences and don't depend on the platform statement to give them meaning. These should begin with a capital and end with a full stop. For example:

> **The speaker made three points:**
>
> - **Aboriginal people across the world have been persecuted in the name of civilisation and religion.**
> - **Even so-called enlightened governments have broken treaties made in good faith by aboriginals.**
> - **Despair among aboriginals will lead either to their cultural disintegration or uprisings against authority.**

This treatment is particularly useful when a listed item is long and detailed, perhaps with several separate sentences, when it would seem odd if the listed item began with a lower-case letter, went on with a new sentence, and ended in a semicolon.

Normally it is not enough to end listed items with a comma or no punctuation at all, though this is often done. The signal is not strong enough: a semicolon or full stop gives more warning of a major break. No standard, however, can apply in every case and there may be times when you need to deviate from it.

Numbering the listed items

There is no need to number the listed items if you or the reader will not need to refer to them again or if you wish to avoid suggesting that the items are in order of priority. Instead, just use a dash followed by a space, or a 'bullet' •.

Other options are arabic numbers (1, 2, 3) or bracketed letters (a), (b), (c) or, as a last resort, bracketed roman numerals (i), (ii), (iii). Roman numerals can be used if you need to put a list within a list – more common in legal documents than in everyday writing:

> **The court may in an order made by it in relation to a regulated agreement include provisions:**
>
> **(a) making the operation of any term of the order conditional on the doing of specified acts by any party;**
> **(b) suspending the operation of any term of the order:**
> **(i) until the court subsequently directs; or**
> **(ii) until the occurrence of a specified act or omission.**

7 Negative to positive

Guideline: *Put your points positively when you can.*

'Accentuate the positive, eliminate the negative,' went the old song. In writing, negatives include 'un-' words like 'unnecessary' and 'unless'; verbs with negative associations like 'avoid' and 'cease'; as well as the obvious ones like 'not', 'no', 'except', 'less than', 'not less than' and 'not more than'. When readers are faced with a negative, they must first imagine the positive alternative then mentally cancel it out. So when a newspaper declares:

> **It is surely less painful to be unemployed if one is not sober, drug-free and filled with a desire to work**

readers have to work very hard to assemble the meaning.

A single negative is unlikely to cause problems, though many a voter has paused, pen poised, when confronted with the polling booth challenge:

> **Vote for not more than one candidate**

instead of the plainer and positive

> **Vote for one candidate only.**

But when two, three or more negatives are gathered together in the same sentence, meaning may become obscure, as in this note from a lawyer to his client, an underwriter:

> **Underwriters are, we consider, free to form the view that James Brothers have <u>not yet proved</u> to their satisfaction that the short-landed bags were <u>not discharged</u> from the ship, and were <u>not lost</u> in transit between Antwerp and Dieppe, when they were <u>not covered</u> by this insurance policy**

– a rodeo ride that perhaps only the lawyer who created it could complete successfully. The going is a little easier here in a pension contract:

'Dependant relative' includes a member's child or adopted child who has <u>not attained</u> the age of 18 or has <u>not ceased</u> to receive full-time education or training.

Put positively this would say:

'Dependant relative' includes a member's child or adopted child who is <u>aged 17 or under</u> or <u>is in</u> full-time education or training.

Just as 'at least' is an ever-present help in a document full of not-less-thans, the word 'only' – which is positive but restrictive – is a useful converter of negative to positive:

The government will not consent to an application if those with a legal interest in the common land object to the application, except in exceptional circumstances.

'Not' and 'unless' both vanish under the influence of 'only', though you may feel there has been a change of emphasis:

Only in exceptional circumstances will the government consent to an application if those with a legal interest in the common land object to the application.

or

If those with a legal interest in the common land object to an application, the government will consent to it only in exceptional circumstances.

Negatives are, of course, useful. Many commands are more powerful in the negative, which is why they have a place in procedures and instruction manuals. Even a double negative like:

Do not switch on the power unless you have made all the necessary checks

is probably more forceful than:

Only switch on the power when you have made the necessary checks.

But avoid overusing negatives, and make sure that those you use are necessary.

8 Cross-references, cross readers

Guideline: *Reduce cross-references to a minimum*

The government has just sent me a 26-page form that includes 183 questions, 70 separate paragraphs of notes and 125 cross-references from questions to notes. I am whizzed from point to point like a pinball from the flippers.

In any complex document some cross-references are inevitable but they should be kept to the minimum. Often, in forms, notes can be positioned next to questions so that readers can find them easily. Unless cross-references are carefully controlled, this kind of horror from a pension policy is the likely result:

> **In the event of the policyholder being alive on the vesting date and having given (or being deemed to have given) appropriate notice in accordance with provision 11.4 the provisions set out in this provision 5 shall apply provided that where by reason of the policyholder's exercise of the option under provision 6.2 or 6.3 the vesting date is a day which is not the specified date, provision 5 shall apply subject to any consequential alterations arising under the relevant part of provision 6.**

Plain English can't improve this very much. The only hope is to scrap it and start again. If the point can't be made in a way that people can understand, it may not be worth making at all: the scheme underlying the policy might need to be simplified.

Skeins of cross-references are not always so hopelessly tangled. Here's an example from an investment policy:

> **If you choose to receive income payments from your investment within the policy, subject as provided in clause 12, income (including tax credits) will be paid to you (subject to such sums being available first to pay any sums due to Unicorn Unit Trusts Limited under clause 9) monthly or quarterly and at any level you choose between a**

minimum of 5 per cent and a maximum of 10 per cent (in increments of 0.5 per cent) based on either a percentage of your original investment or the value of your policy at the time you choose to start making withdrawals.

To isolate and jot down the three main points in this 98-word sentence is to go a long way towards clearing up the mess:

1 If you have chosen to draw income from the investment, we will pay it to you, monthly or quarterly.

2 How much we pay you could be affected by clause 12 and by Unicorn's charges set out in clause 9.

3 We will pay you whatever percentage you choose – within certain limits – of your original investment or of the value of your investment at the time you start to make withdrawals.

By putting these points in 1-3-2 order, the cross-references can be grouped into a separate sentence at the end. The result might say this, using short sentences and paragraphs:

If you choose to draw income payments from your investment in the policy, we will pay them to you monthly or quarterly. The payments will include tax credits.

The payments you choose must be a percentage of your original investment or of the value of your investment at the time you start receiving the income. Your choice of percentage must be at least 5 per cent but not more than 10 per cent (using 0.5 per cent steps).

Payments are subject to clause 12 and are available first to pay any money owed to Unicorn Unit Trusts Limited under clause 9.

This is still not simple because the ideas themselves are complicated and, though there are fewer words, they take more space. But now readers have a much better chance of understanding the points.

9 Clearly non-sexist

Guideline: *Try to avoid sexist usage.*

A poster on the London underground is trying to sell cat litter. Not easy, perhaps, but since many cats are owned by women, the task becomes harder if the poster starts by excluding them:

> **Keep your fur on. Your owner is as concerned with hygiene as you are, which is why he buys Catsan Hygiene Cat Litter... In fact, Catsan works so well that it actually produces less odour after six days than ordinary cat litters do after only one day... Which means your owner will have to change it less often. Allowing them to get on with their life and you to get on with yours, all nine of them.**

Though sexist usage is not strictly a matter of clarity, any writing habit that builds a barrier between you and half your readers must reduce the impact of your message. So even if you disagree with the view that sexist writing reinforces prejudice and discrimination, it is still wiser to use inclusive language. Occasional silliness apart ('the art of one-upping' doesn't have quite the same ring as 'one-upmanship'), inclusive writing usually makes more sense and is more accurate.

Using sex-neutral terms

Using sex-neutral terms means avoiding words which suggest that maleness is the norm or superior or positive and that femaleness is non-standard, subordinate or negative. Sex-specific terms like 'businessmen', 'firemen', 'poetess', 'headmaster', and 'sculptress' can be replaced by less restrictive words: 'business people' (or 'executives'), 'firefighters', 'poet', 'headteacher', and 'sculptor'. Some sex-specific terms may survive this trend, such as 'actress' – which perhaps lacks some of the negative echoes of low status and low pay. (If you were writing about acting, you could always ask some female actors for their preference.) 'Fishers' and 'fisherfolk' seem unlikely to gain popular acceptance as alternatives to 'fishermen' on the open sea, yet 'anglers' is a convenient sex-neutral term for those who fish for sport. There seem not to be any palatable alternatives for such terms as 'manned space flight' and 'craftsmanship'.

Sex-specific words	*Sex-neutral words*
authoress	author
chairman, chairwoman	chair (or ask what they prefer)
clergymen	clergy, clerics
craftsmen	craft workers, artisans
executrix	executor
forelady, foreman	supervisor, head juror (law)
heroine	hero
layman	lay person
man, mankind (humans)	the human species, human beings, people, humans
man (noun)	person, individual, you
man (verb)	operate, staff, run, work
manageress	manager
man-hours	working hours, work-hours
man-made	manufactured, artificial
policeman, policewoman	police officer
salesman, salesgirl	sales agent/representative/assistant
testatrix	testator, will-maker
workman	worker

Using titles or 'he' and 'she'

It is better to avoid 'he', 'his' or 'him' when you intend to include both men and women. Instead of:

> **Solvent abuse is not a crime but if a police officer finds a young person under 17 sniffing solvents, he should take him to a secure place such as the police station, home or hospital.**

you could repeat the short titles of both people:

> **Solvent abuse is not a crime but if a police officer finds a young**

person under 17 sniffing solvents, the officer should take the person to a secure place such as the police station, home or hospital.

Using 'he or she' and 'him or her' is also feasible here – perhaps, to avoid confusion, for just one of the people:

Solvent abuse is not a crime but if a police officer finds a person under 17 sniffing solvents, he or she should take the person to a secure place such as the police station, home or hospital.

Repeated use of 'he or she' and similar terms becomes clumsy and obtrusive. The alternatives, 's/he' or 'he/she', look ugly and cannot be spoken easily.

In guidance to police officers, it would be feasible to use 'you':

Solvent abuse is not a crime but if you find a person under 17 sniffing solvents, you should take him or her to a secure place such as the police station, home or hospital.

Using the plural

A further alternative, and often the best, is to use the plural:

Solvent abuse is not a crime but if police officers find a person under 17 sniffing solvents, they should take the person to a secure place such as the police station, home or hospital.

Using plurals as singulars

It is becoming more acceptable to flout the grammatical conventions set in the eighteenth century by male grammarians and to do what Shakespeare did when he wrote:

God send everyone their heart's desire.

In other words, to revive the old use of 'they', 'them' and 'their' as singulars:

Give details of your partner's income. If they have been unemployed for more than 12 months . . .

You may find that an individual has levels of competence in several skills beyond those required in their current role. This will occur when

someone has developed their skills and potential in readiness for other opportunities.

To me, these read smoothly enough, though the second may be better in the plural:

You may find that individuals have levels of competence in several skills beyond those required in their current roles. This will occur when they have developed their skills and potential in readiness for other opportunities.

Postscript: the alternatives are worse

To reject all the ideas as bad compromises means accepting the kind of writing found in this advert from the 1920s, in which all typists are assumed to be women and all managers men:

By dictating . . . you enable your typist to be typing all day, not wasting half the morning taking notes . . . She writes more letters and does her work more easily and accurately. Also the Dictaphone saves the time of the chief. We hardly like to talk about the 'after-hours' work, but it is acknowledged that closing time is the time when men who bear the real burdens oftentimes get down to their serious work. Then is the time when a man can concentrate, when there are no interruptions, no clicking of typewriters, no buzz of conversation to disturb his train of thought. Minds become decisive, delicate situations are mastered, and tactful, forceful phrases find shape. Here it is that the Dictaphone comes into its own; it records the accumulated letters for the typist first thing in the morning, the chief's desk is clear – his mind relieved, and he arrives next day with only the 'current' matter to receive his attention.

10 Sound starts and excellent endings

Guideline: *In letters, avoid fusty first sentences and formula finishes.*

In letters and memos the best place to start being plain is at the beginning. Your first sentence should be clear, complete, concise and written in modern English. This is easily done if you steer clear of three common traps.

Trap 1: Writing half a sentence instead of a full one

All these first sentences are incomplete:

> **Further to your letter of 3 February concerning the trustees of the P F Smith 1982 Settlement.**
>
> **With reference to our previous correspondence notifying you of the transfer of the administration of the above policy to Norwich Union Healthcare with effect from 9 January.**
>
> **In response to your letter to Mr Jones dated 19 February.**
>
> **Regarding your claim for attendance allowance.**
>
> **Referring to your letter dated 10 July about delayed frequency allocations.**

Each statement needs to be continued and completed, inserting a comma instead of a full stop. For example:

> **Further to your letter of 3 February concerning the trustees of the P F Smith 1982 Settlement, I am pleased to enclose the form you requested.**
>
> **Regarding your claim for attendance allowance, I need to ask you for some more information.**

Alternatively, you could insert a main verb at the start to complete the sentences:

Thank you for your letter of 3 February concerning the trustees of the P F Smith 1982 Settlement.

I refer to your claim for attendance allowance.

Other verbs that will do a similar job include 'I acknowledge'; 'I confirm'; 'I write to explain'. These alternatives are preferable for another reason: they use personal words like 'I', 'you' and 'we'.

Don't be afraid to write a one-sentence paragraph at the start of a letter:

Thank you for your letter dated 13 May.

If you are taking the initiative – rather than responding to someone's enquiry – these phrases may be helpful to get your first sentence off to a sound start:

You are warmly invited to . . .

You may be interested in . . .

This is an opportunity to . . .

Now is a good time to consider . . .

Or you could ask a question – preferably one to which the reader will answer yes:

Does your office have old and outdated law books gathering dust in corners? Would you like to create some extra space for yourself, and see those books go to a good cause at the same time? If so . . .

Trap 2: Repeating the heading

Most business and official letters benefit from a heading, as it introduces the topic and saves having to write a long first sentence to cover the same ground. The heading (usually underlined or in bold type) might even run to several lines:

Curtis Brothers Ltd
Lease Agreement No. 727-252-8978
Goods: Touchtone Telephone System
Installation address: Piller House, Crook Street, Downtown

Whether the heading is short or long, don't keep referring to it with such phrases as 'with reference to the above-mentioned equipment', 'regarding the above-numbered agreement', 'the above matter' and 'in connection with the aforementioned'. Let's say you are selling property and someone has phoned for details of a house. Instead of writing:

Dear Ms Widdicombe

'The Tree House', 67 Larch Avenue, Poplar

I refer to your enquiry yesterday in relation to the above-mentioned property, and enclose details as requested

it is crisper to write

Dear Ms Widdicombe

'The Tree House', 67 Larch Avenue, Poplar

I refer to your enquiry yesterday and enclose details of the property as requested

or

Thank you for your enquiry yesterday about this property; details are enclosed

or, rather more syrupy

Your enquiry about this property is much appreciated. I am pleased to enclose details.

This last suggestion has the advantage of mentioning the reader first by using the phrase 'your enquiry' – often a good idea since readers are usually more interested in themselves than in you.

Trap 3: Archaic usage

Avoid such fusty phrases as:

Our recent telephone discussion refers.

Modern readers find 'refers' very odd in this position. Use 'I refer to . . .' or 'In our recent telephone conversation, we discussed . . .'. In an internal document you could merely say 'We spoke about this.', when the heading will provide enough context for what you spoke about.

Re your letter of the 10th, the contents of which have been noted.

'Re' is Latin for 'in the matter of' and is best avoided. When responding to someone's letter it is stating the obvious to say that you have 'noted the contents'. And anyway, the phrase sounds bureaucratic and disapproving – avoid it.

Thank you for your letter of the 15th instant.

Avoid 'instant' and 'ultimo', which some lawyers still use to mean 'this month' and 'last month' respectively.

Finishing well

Keep the finish simple, warm and sincere. In a sales letter you could reiterate an important action point:

I look forward to receiving your application soon.

You could even add a PS beneath the signature. Useful non-selling sign-offs include 'I hope this is helpful' (which is much better than the negative and servile 'I am sorry I cannot be more helpful'); and 'If you need more information, please contact me' (which is better than waffle like 'Should you require any further assistance or clarification from ourselves with regard to this matter, please do not hesitate to contact the writer undersigned').

If you are trying to close a correspondence with a difficult customer, you could try:

I hope you will understand our position and I regret that we cannot help any further.

Remember: these are only suggestions. Compose your own phrases to suit your own circumstances and personality.

Conventions on opening and closing

If you start with 'Dear [first name]', 'Dear [Mr/Mrs/Miss/Ms]', or 'Dear [first name and surname]', the conventional endings are 'Yours sincerely', 'Sincerely yours' or 'Yours truly'. Alternatively, use a sensible sign-off of your own making: 'Kind regards' and 'Best wishes' are now common among people well known to each other.

If you begin 'Dear Sir' or 'Dear Madam', the traditional formal ending is 'Yours faithfully'. An alternative opening is to use the organization's name: 'Dear Express Bus Company.'

Esquires were originally the shield-bearers of knights, or men in royal or noble service. Once a favoured courtesy title after a man's name, 'Esq' is now rarely seen in the address block of a letter. It is perhaps thought over-formal, silly or servile. The title is never applied to a woman's name. Nor does it traditionally follow 'Mr', so 'Mr J Smith Esq' is unusual.

Punctuation can be slashed from the address block, so this:

J. H. Author, Esq.,
34, Ivory Lane,
Kelvintown,
KT2 0NN.

becomes:

Mr J H Author
34 Ivory Lane
Kelvintown
KT2 0NN

11 Using good punctuation

Guideline: *Put accurate punctuation at the heart of your writing.*

Punctuation shouldn't cause as much fear as it does. Only about a dozen marks need to be mastered and the guidelines are fairly simple. What's more, you can see the marks being well applied every day in the serious newspapers, where 99 per cent of the punctuation will accord with the advice given here. Study of press punctuation also reveals the full range of marks, as journalists – especially the star columnists – use far more colons, semicolons and dashes than business and official writers.

The rumour has got about that any old punctuation, or none at all, will do. Not for a successful plain English style it won't.

A good command of punctuation helps you to say more, say it more interestingly and be understood at first reading. Punctuation is an essential part of the tool-kit – as important as choosing the right words.

Consider these statements:

Mother to be attacked on waste land.

Never mind people who dislike cats are in a minority.

Once she had the dress off she would go in search of matching shoes, gloves and a handbag.

Without good punctuation they can be interpreted in different ways or make no sense at all, distracting the readers and causing them to stop and backtrack. Punctuation helps to fix the meaning and smooth the path:

Mother-to-be attacked on waste land.

Never mind: people who dislike cats are in a minority.

Once she had the dress, off she would go in search of matching shoes, gloves and a handbag.

Punctuation shows how words and strings of words are related, separated and emphasized, so its main purpose should be to help the reader understand the construction of the sentence. A lesser purpose is to act as a substitute for the devices we all use in speech, such as pausing and altering pitch; but an idea such as 'use a comma when you would have a one-beat pause in speech, a semicolon for a two-beat pause, and a full stop for a three-beat pause' is useless because it can be interpreted in too many ways. Even so, you might want to read a sentence aloud to help you decide how to punctuate it.

Though this chapter describes the standards that most careful writers accept as sound, no two writers will ever agree on the position of every comma and full stop. So if you are editing someone's writing, be prepared for a bit of give and take.

Today, even careful writers have to face the fact that fine distinctions between such marks as colons and semicolons will be lost on many of their readers. Yet so long as readers regard a semicolon as a funny-looking comma and a colon as a bit like a full stop, that will usually give them enough to hang onto. So, even if you decide to use the less common punctuation marks sparingly when writing for a mass audience, there is no need to remove them altogether.

Full stop (.)

As shown in chapter 1, the main use of a full stop (or, in the US, *period*) is to show where a sentence ends. For this reason, full stops should be the most common mark on the page.

There's no need to use full stops in people's names or in abbreviations or acronyms – *Mr J C Bennett, BBC, US, eg, ie, 8am, 9pm* – unless there's a genuine chance of ambiguity. Headings in a report, letter or memo don't need a final full stop.

Comma (,)

Single commas act as separators between parts of a sentence:

> **Using accident and ill-health data and drawing upon my experience of inspecting slaughterhouses and meat-processing plants, I have prepared a list of hazards found in the meat industry.**
>
> Or:
>
> **Although suitable protective equipment was available, most of the operatives were not wearing it.**

Be sparing with commas. Putting them in every few words prevents the reader getting the construction of the sentence. This, for example, is a mess:

I have tried on a number of occasions to contact you at your office, without success and now resort, to writing to you, as these items should be dealt with urgently. You will appreciate the possible problems involved, if these units become occupied, before we have had a chance to check for their compliance with standards, and I would be obliged, if you could respond, by complying with my inspector's original request for compliance, as any further delay, could result in the start of disciplinary action against your company.

Almost every comma could go and the long final sentence could be split.

A pair of commas cordons off information that is an aside, explanation or addition. Readers can, if they wish, leapfrog the cordoned-off area and still make sense of what is said:

Holmes, having searched for further clues, left by the back door.

Sometimes the position of commas will change the meaning, so be careful with them:

The girls, who will join the team next week, are fine players for their age.

Here, the information between the commas comments on 'the girls', but is not essential to the main point of the sentence, which is that the girls are fine players. This kind of insert is said to be a commenting clause. Compare this, without commas:

The girls who will join the team next week are fine players for their age.

Obviously the meaning has changed: the words 'who will join the team next week' now help to define the girls. They are essential to our understanding of what kind of people they are. The clause is said to be defining and does not need a cordon of commas.

Commas are helpful in separating listed items:

Staple foods include rice, wheat, sorghum and millet.

A comma can also create special effects, such as suspense. Compare this:

They crept into the room and found the body.

with

They crept into the room, and found the body.

Finally, ignore the advice often heard on writing courses: 'Use a maximum of one comma per sentence.' It is just plain silly.

Colon (:)

Colons have three main purposes:

1 To introduce a vertical list (as in the line above) or a running-text list such as:

She has several positive characteristics: charm, dignity and stickability.

2 To act as a drum roll or a 'why-because' marker which leads the reader from one idea to its consequence or logical continuation, for example:

There's one big problem with tennis on radio: you can't see it.

3 To separate two sharply contrasting and parallel statements:

During Wimbledon, television is like someone with a reserved ticket: radio is the enthusiast who has queued all night to get in.

A weaker contrast might be signified by a semicolon; there is some overlap in meaning between the two marks.

In all these uses, the colon will usually follow a statement that could be a complete sentence. After the colon the sentence will usually continue with a lower-case letter.

A colon does not need support from a dash (:–). This nameless thing is not acceptable to most publishers, wastes a key stroke, and looks repulsive in most word-processed documents.

Semicolon (;)

Consider these sentences from a notice next to an ancient cathedral clock:

The large oak frame houses the striking train of gears, these parts have been painted black and are the early parts of the clock. In this

same frame on the left between two posts is a going train (a time piece), this has been painted green.

Both commas are wrong and should be semicolons or full stops. To use semicolons safely – ignoring for the moment their use in lists – you need to satisfy two criteria:

1. The statements separated by the semicolons could stand alone as separate sentences.
2. The topics mentioned in the two statements are closely related.

So the paragraph would say:

The large oak frame houses the striking train of gears; these parts have been painted black and are the early parts of the clock. In this same frame on the left between two posts is a going train (a time piece); this has been painted green.

A semicolon can often seem less curt than a full stop. Instead of starting a letter:

Thank you for your letter of 10 December. We apologize for the delay in replying.

it would be better to write:

Thank you for your letter of 10 December; we apologize for the delay in replying.

A comma would be wrong in that position because it cannot sustain such a long pause. Rarely is a comma enough to separate complete sentences.

Semicolons can also be used instead of commas to separate items in a list:

Lunch at Henry's comprised all the ingredients for a happy and contented life: wine from the grapes of Tuscany; tropical avocados, seductively soft and yielding; French king prawns, clothed in a luscious sauce; a multitude of meaty snacks for carnivores and nutty nibbles for vegetarians; and the company of all the beautiful people of the town.

In a long, catalogue-type list, semicolons are ideal dividers:

Target audiences for the new manual will include other companies in our group, both European and US-based; business leaders, top

politicians and other leading opinion-formers; consultants of proven expertise; and local schools and colleges.

To use merely commas as dividers would produce chaos because commas already exist within some of the listed items. An introductory colon after *include* would be unnecessary, though harmless.

Dash (–)

Dashes are sometimes used singly to indicate the start of an aside, explanation or addition:

Justifying their case, smokers introduce a herring so red that it glows like coal: that if their illnesses are self-inflicted, well, so are most people's – look at traffic accidents, look at potholers.

They can add emphasis, too:

He shot big game for status, pleasure – and greed.

When used in pairs, dashes draw special attention to the phrase they surround (compare 'Brackets', below):

Visitors may stay overnight – or for as long as they wish – in the hostelry run by the friars.

A pair of commas or a pair of brackets would have done just as well, but the long dashes emphasize the point.

It is all right to use more than one pair of dashes in a sentence but take care that the meaning doesn't disintegrate. Here's an awkward, if playful, use of two pairs of dashes by a newspaper columnist:

Volkswagen is in trouble – terrible trouble – very terrible trouble, and we can sit on the sidelines – entry free – and bask in somebody else's trouble for hours on end.

Perhaps he should have read his paper's own guide on the subject of dashes:

Dashes are sloppy punctuation, ugly in narrow columns of newspaper type. They often indicate that a sentence is badly constructed and needs rewriting.

Happily this advice is ignored by everyone on the paper.

In typewriting, where dashes are not usually available, it is customary to use a hyphen with a space either side. In typesetting or desk-top publishing, most publishers use a spaced 'en rule' – which is what I have just used.

Brackets ( )

Brackets (also called round brackets or, in the US, parentheses) surround an aside, explanation or addition that is relatively unimportant to the main text (compare 'Dash', above):

> **The biker fell off at 160mph, suffering an open fracture of his tibia (lower leg) and losing two inches of bone in the process. Only ten years ago, four out of ten open tibial fractures (in which the bone comes partially out of the leg) resulted in amputation.**

Well-placed brackets can prevent the meaning of a sentence disintegrating, even if the result looks cluttered:

> **The problem is tidal flooding along the river. Many plans have been studied but the agreed solution is a system of river mattressing (to avoid breaches) and embankment raising (to avoid exceptionally high-tide flooding).**

If a sentence begins within a bracket, it should start with a capital and end with a full stop inside the bracket. A comma should rarely precede a bracket but there is no harm in putting one after the closing bracket if this aids meaning.

If you are going to use an acronym several times and the reader is unlikely to know what it means, spell it out on the first occasion and put the acronym in brackets:

> **Agency Services Limited (ASL) has tendered for the work.**

Square brackets [ ]

Square brackets show that the text within does not belong to the document or quotation but is being inserted for clarity:

> **He [Mr Smith] told me to go home.**

Capitals

Capital letters (upper case) defy any would-be rule-maker, as some decisions on whether to dignify a word with an initial capital have to be left to the individual writer. Use them sparingly though, and err on the side of lower case. Writing afflicted with random capitalitis just looks silly:

> **Please Complete One Line For Each Person Normally Living At This Address, including those who are temporarily living away. Include Everybody – Yourself, All Adults, whether entered or not on the Electoral Registration form, All Younger Persons, Children and Babies.**

Headings in business and scientific reports tend to suffer from the same disease, with writers making invidious distinctions between words they capitalize and words they don't. Compare the distracting:

> **Factors that May Affect the Success of the Strategy**

with

> **Factors that may affect the success of the strategy**

People: Use an initial capital for ranks and titles when attached to a person's name, thus *Prince Charles* but *the prince*, *President Clinton* but *the president*, *Judge Wright* but *the judge*. Offices of state in the UK such as *home secretary* and *chancellor of the exchequer* are lower case in *The Economist* and *The Sunday Times*, but become *Home Secretary* and *Chancellor of the Exchequer* in *The Times*. I prefer lower case. Titles which look odd in lower case need capitalizing, hence *Master of the Rolls*, *Lord Chief Justice*. In some organizations, job titles like *managing director* are given initial capitals but this is poor practice – at what rank do you stop?

Organizations, acts of Parliament and government departments generally take initial capitals only when their full name or something similar is used. Thus *Metropolitan Police* but *the police*; *County Court* and *Court of Appeal* but *the court; the Health and Safety at Work etc Act* but *the act; Legal Aid Fund* but *legal aid*. A newspaper would print *Blankshire District Council* but *the council*. The council itself, however, might prefer to use a capital c whenever it referred to itself.

Trade names take upper case: *Hoover* but *vacuum cleaner*, *Xerox* but *photocopier*, *Penbritin* but *ampicillin*.

The names of wars and historical periods normally take an initial capital: *Iron Age fort*, *Vietnam War*.

Acronyms that can easily be spoken as words need only have an initial capital: *Nato, Unicef, Unesco*. Otherwise, use all capitals: *BBC*, *EU*, *USA*.

Hyphen (-)

Hyphens make links. For example, they link words which form a composite adjective before a noun: *computer-based work, short-term goals, half-inch nail, three-year-old child, out-of-hours work, no-go area, couldn't-care-less attitude*. This kind of hyphenation helps to reduce ambiguities and stumbles. A guest house says:

> **The new owners of this beautiful mansion take pride in their home cooked meals.**

When the readers reach *home*, they stop momentarily and think 'that's the end of the phrase', only to be caught out by the next word; *home-cooked* needs a hyphen.

A company writes:

> **Normally on oil and gas related enquiries, a briefing meeting is convened at the client's request.**

Gas related comes before *enquiries* and is effectively one word, so it needs a hyphen. But there are oil-related enquiries too in the sentence, because *related* is implied after *oil*. So the sentence should say, with a hyphen hanging after *oil*:

> **Normally on oil- and gas-related enquiries, a briefing meeting is convened at the client's request**

or

> **Normally on oil-related and gas-related enquiries, a briefing meeting is convened at the client's request.**

It is difficult to use hyphens consistently, and there is usually someone to disagree with your best efforts. The treasurer of the World Pheasant Association might be peering over your shoulder:

> **[Your] article . . . refers to the endangered 'white-eared pheasant' . . . The hyphen is incorrect. The pheasant in question is a white pheasant with ears. There are also blue eared pheasants and brown eared pheasants. The taxonomic similarity is that all the three pheasants have ears and all the ears are white.**

When a woman was assaulted by a London taxi driver a newspaper described the attacker as a *black cab driver*. But which was black, the driver or the cab? The paper had meant to say *black-cab driver*, leaving his skin colour out of the question. Had it really wanted to mention his colour, it could have written *black black-cab driver* or *white black-cab driver*, but perhaps then it would have decided to recast the sentence.

Some nouns formed by two or more words need hyphens: *run-up, build-up, punch-up*. Without a hyphen, there is ambiguity in sentences like:

Last week there was a bust up in the street

where the possible meanings include fight, drug bust or piece of sculpture.

You rarely need to hyphenate a verb – all three hyphens in this advertising leaflet are unnecessary:

With inflation decreasing, the press and politicians are spelling-out the message that the recession is over . . . Now is the time to think about making cash available for the numerous high street bargains and to begin paying-off some of the commitments that have built-up in this tough period.

You might, however, write of the need to power-up a machine. To leave the words unjoined would seem odd.

It's normally unnecessary to hyphenate when one of the words is an adverb (see chapter 13):

rapidly growing economy; carefully crafted answer

though some newspapers routinely do this.

Apostrophe (')

The apostrophe is now so widely misused – an errant tadpole, one columnist calls it – that its eventual death seems inevitable. Alongside their no-smoking stickers, companies could soon be declaring themselves apostrophe-free zones. This would be a pity, as the correct use of apostrophes conveys meaning and prevents ambiguity, while misplaced apostrophes make the reader stumble and backtrack.

Use an apostrophe to show possession. Normally, add *'s* to the person or thing possessing:

The judge's wig was eaten by the general's horse.

So the wig belongs to one judge and the horse to one general.

The people's leader ignored the children's opinions.

Here the leader belongs to the people (not the people*s*) and the opinions belong to the children (not the children*s*).

Add only an apostrophe if the things or people possessing already end in *s*:

The three inspectors' reports coincided.

Several players' cars were broken into.

When the singular form of a word ends in *s*, either add *'s* or just an apostrophe. Both these forms are correct:

I sent you Mr Jones's copy of the lease yesterday.

I sent you Mr Jones' copy of the lease yesterday.

Words that end in *ss*, like *boss* or *business*, also take *'s* in possession:

The boss's failure to act led to orders being lost.

Use an apostrophe to show that a letter is missing:

Today's the day for a fresh start (Today is . . .)

It's no concern of mine (It is . . .)

Doesn't anybody have the results? (Does not . . .)

This surreal, apostrophe-free message is to be seen roaming Britain on a fleet of lorries:

COLLECTING TOMORROWS
DELIVERIES TODAY

As the deliveries belong to tomorrow, there should be an apostrophe between the *W* and *S*.

Remember that pronouns – words in place of nouns – like *his, hers, ours, yours, theirs* and *its* don't need apostrophes. They are words in their own right – no letters are missing and the possession is built into the word. (The only exception is *one's*, as in *one's head is aching*.) So it would be correct to write:

The cat wandered in, its wet paws patterning the carpet.

Though the wet paws belong to the cat, there is no apostrophe in *its* because the possession is built into the word. Only insert an apostrophe in *its* when the word is short for *it is*.

In some expressions of measurement and time, apostrophes are conventional:

They took a week's holiday without permission and the firm then gave them two weeks' notice.

It is difficult to see the logic in this usage, but in the first example *holiday* in a sense belongs to the *week*. Without the apostrophe it would say:

a weeks holiday

which would mix a singular *a* with a plural-looking *weeks*. So, if it's logical to write *a week's holiday*, you should move the apostrophe along one letter when the possessor ends in *s*, as in *two weeks' notice*; this, however, is becoming extinct.

Generally, don't use an apostrophe in a plural:

In the 1960s, even some MPs wore their hair long

unless there is possession:

Six MPs' offices were ransacked.

Refuse to be led astray by any shop notice proclaiming:

Shrimp's, prawns', pears', orange's and sandwich'es.

These are all straightforward plurals – no apostrophe is needed. Sometimes an exception to the rule is demanded by the need to avoid ambiguity. Hence:

Mind your p's and q's

A list of do's and don'ts

The first animal in the dictionary has three a's in its name

are preferable to:

Mind your ps and qs

A list of dos and don'ts

The first animal in the dictionary has three as in its name.

Occasionally you meet a rarity where at first the apostrophe looks superfluous:

Maria's lasagne was better than the chef's.

Clearly the lasagne belongs to Maria, so the first apostrophe is fine. But what belongs to the chef to deserve an apostrophe? Lasagne, again, because that word is implied. Had there been more than one chef, the apostrophe would have followed the s. The same thing happens with some shop names. *Lewis's*, in Liverpool, means the shop belonging to Lewis.

In some cases, an apostrophe might seem appropriate but is unnecessary. If you call your office guidance on word processing a *Typists Handbook*, you could argue that it's a handbook *for* typists not *belonging* to them, so an apostrophe is not needed. This is perhaps why the National Association of Citizens Advice Bureaux has no apostrophe in its name – the bureaux do not *belong* to citizens, they are *for* citizens to use.

A final point: many keyboards hide apostrophes and curly quotation marks but seductively offer instant access to the marks for feet (') and inches ("). Seek out the proper marks – they'll usually be hidden under another key.

Ellipsis (. . .)

This has two purposes:

1 To show that material is missing, perhaps from a quotation.
2 To indicate suspense:

He said: 'And the winner is . . . Sydney, Australia!'

There should be three dots in the ellipsis, not two, five or seven. In typesetting or desk-top publishing, there should be a word space before and after each one, so . . . not... Some typesetters insist on adding a fourth dot when a sentence ends with an ellipsis, the fourth dot being set tight on the third to show its different character.

Quotation marks (' ')

These, also called quotes, speech marks or inverted commas, indicate the opening and closing of direct speech:

'There is no alternative,' said the prime minister, 'so we will go on as before.'

Or, if the sentence ended after the first statement:

'There is no alternative,' said the prime minister. 'We will just go on and on as before.'

Most books and newspapers use single quotes, reserving double quotes for a quotation within a quotation. The reverse is acceptable, though.

Quotes are also used to clothe a word in irony or, usually unnecessarily, to apologize for some clumsy or supposedly colloquial usage:

Recently I have been 'toying' with the idea of buying a mobile phone and wondered if you were going to sell them.

In this preliminary phase I also 'touched base' with many other interested people.

Exclamation mark (!)

Use this to follow exclamations of surprise, shock or dismay, but sparingly and singly!!! Using the mark to signal a witticism or acute observation not only deprives readers of the pleasure of discovering it for themselves, but suggests they are too slow to perceive it without help.

Question mark (?)

Use this after direct questions. There is no need for one if the question is really a polite demand:

Will you please let me have your reply by 6pm today.

12 Seven writing myths explored and exploded

Guideline: *Avoid being enslaved by writing myths.*

A business writer once told me that her English teacher had ordered her never to begin consecutive paragraphs with the same letter of the alphabet. After 30 years of following this non-rule, she told me she was beginning to wonder if there was any justification for it.

Your path as a writer is steep enough without the extra burden of schoolroom mythology. This chapter examines some of the most tenacious non-rules imposed on writers.

Myth 1: You must not start a sentence with 'but'

This ban is even extended to 'so', 'because', 'and', and 'however'. It is not a rule of grammar, and may originate from a desire among teachers to persuade children to link up the sentence fragments they tend to write.

Most good authors over the last few hundred years have ignored the myth. Jane Austen begins sentences with 'but' on almost every page, and occasionally uses 'and' in the same position, in the sense of 'furthermore'. This is one of her buts from *Mansfield Park* (1814), where it prevents the bland first sentence from running on too long and heightens its contrast with the ironic second:

> **She had two sisters to be benefited by her elevation [marriage to a social superior]; and such of their acquaintance as thought Miss Ward and Miss Frances quite as handsome as Miss Maria, did not scruple to predict their marrying with almost equal advantage. But there certainly are not so many men of large fortune in the world, as there are pretty women to deserve them.**

'But', like most sentence connectors, signals a shift in pace and direction. It may help in stating an argument or point of view, as in this example from a modern journalist who, for emphasis, happily begins the final sentence with 'And':

> **The children of MPs, royalty, journalists and other moral prodnoses do not need to read underclass horror-stories to find out about the lifestyle problems which adult sexuality inflicts on children. They are familiar with it all: access arrangements, vendettas, embarrassment, lawsuits, confusion, hypocrisy. 'I believe strenuously,' says Mrs Nicholson [a British MP], 'that every child deserves a mother and father'; and so say all of us. But the plain fact is that not every child has them to hand. And in family life, the golden rule is to start from where you are.**

Even old-time grammarians felt able to start sentences with 'but'. In his *Manual of English Grammar and Composition* (1915), J C Nesfield says:

> **. . . it is convenient for the sake of brevity to say that 'a conjunction joins words to words, and sentences to sentences.' But this is not enough for the purposes of definition.**

In short, you can start a sentence with any word you want, so long as the sentence hangs together as a complete statement.

Myth 2: You must not put a comma before 'and'

Many folk insist that putting a comma before 'and' is bad. Ignore them. (Just as Nesfield did, in the second line of the previous example.) Though a comma is usually unnecessary in this position, it may help the reader to see how the sentence is constructed, or put a pause exactly where you want it. A few examples:

> **It was a bright cold day in April, and the clocks were striking thirteen.** (The first line of Orwell's *Nineteen Eighty-four*.)

> **Parents shop around. They send for a clutch of prospectuses, see what's available in both the public and private sectors, and are prepared to switch their children between the two at different stages in their education.** (*The Times*, 1994.)

> **When we met last night, you explained that you no longer wished me to remain as Secretary of State for Education, and I am writing to say how glad I have been to serve in Her Majesty's Government.** (Letter of resignation to the British prime minister, 1994.)

And here is Beatrix Potter, setting a bad example to children in *The Tale of Peter Rabbit*:

> **He found a door in a wall; but it was locked, and there was no room for a fat little rabbit to squeeze underneath.**

Myth 3: You must not end a sentence with a preposition

A few fossils believe that a sentence is bad if it ends in a preposition. According to them, these are ungrammatical:

> **There are certain values we must all be prepared to stand up for.**

> **The council decided that this was the right system to invest in.**

It's true that the first could be recast as:

> **There are certain values for which we must all be prepared to stand up.**

but this would still annoy anyone who thinks 'up' is a preposition here (actually it's part of the verb). In my view there is no point changing a sentence that reads well, sounds right and says what the writer wants to say. Sometimes sentences ending in prepositions do need to be recast, but this is because they sound ugly, not because they break a rule. These sentences, for example, come to a feeble end:

> **Get clean photocopies of the forms you want to make changes to. Make sure there is enough white space to mark your alterations on.**

Better to write:

> **Get clean photocopies of the forms you want to change. Make sure there is enough white space for marking your alterations.**

Myth 4: You must not split your infinitives

Splitting the infinitive means putting a word or phrase between 'to' and the verb word, as in:

> **The department wants to <u>more than</u> double its budget.**

> **The passengers were asked to <u>carefully</u> get down from the train.**

If you think a sentence will be more emphatic, clear or rhythmical, split your infinitive – there is no reason in logic or grammar for avoiding it. The examples above seem better split than not. Take care, though, lest the gap between 'to' and the verb word becomes too great, as the reader could lose track of the meaning.

A newspaper editorial says:

> **The most diligent search can find no modern grammarian to pedantically, to dogmatically, to invariably condemn a split infinitive. Rules are created to aid the communication of meaning. In the cause of meaning they can sometimes be broken.**

Though this is excellent, it encourages the idea that to split an infinitive is to break a rule. Yet there is no such rule, merely a superstition that arose in the nineteenth century when grammarians sought to impose Latin rules on English. In Latin, a present-tense infinitive is always a single, unsplittable word.

If you can't bring yourself to split an infinitive, at least allow others the freedom to do so. Byron, for example, with his:

> **To slowly trace the forest's shady scene.**

Phrases like 'I easily won the race' and 'he quickly drove away' are not split infinitives as there is no infinitive present.

Myth 5: You must not write a one-sentence paragraph

If you can say what you want to say in a single sentence that lacks a direct connection with any other sentence, just stop there and go on to a new paragraph; there's no rule against it.

Myth 6: You should write as you speak

Advice widely heard, but take it with a large pinch of salt. True, many writers would benefit from making their writing more conversational, using more personal pronouns and active verbs. But this is not the same as writing as you speak. Most of us don't speak in complete sentences – a transcript of our talk usually reads as gibberish on the page. Plain English is much more than speech transcribed; it is speech organized, worked and refined.

Myth 7: You should test your writing with a readability formula

Several software grammar-checkers claim to measure the readability of writing under such names as the FOG (frequency of gobbledygook) Index and the Flesch Test. Generally such formulas use only two variables to get their results: sentence length and word length. The score is then compared to a reading age level (UK) or grade level (US), and, hey presto, a scientific result is produced. Yet the formulas disregard such things as the use of actives and passives, the way the information is organized, how it looks on the page, and the reader's motivation and level of prior knowledge. They give only the merest hint about how to write a text better, and they encourage the idea that a clear document is one that scores well on a formula. Few serious writers bother to use them. Have fun with formulas by all means – surprise your bosses with an evaluation of their writing, perhaps – but then forget about them.

13 Conquering grammarphobia

Guideline: *You can be a good writer without learning hundreds of grammatical terms.*

Grammarphobia – an irrational terror of grammatical terminology – is common among those who underwent grammar lessons at school and those who did not. The first group might regard grammar as just one more abstract ritual to be endured while the sun shone outside; the second group might see it as a magical key to successful writing, unreasonably withheld from them by malevolent educationists.

A little grammar goes a long way. It is not necessary to know much beyond primary-school grammar in order to write plain English, but it's helpful to know a few terms if only to get the most out of this book. This chapter provides a brief glossary of the grammatical terms I have used. When reading it, remember that many words change their grammatical character depending on their role in a sentence. Just because 'progress' is a noun in this sentence:

We will make progress on the project next week

doesn't stop it becoming a verb in this one:

I have progressed further than expected.

Adjective A word of description. In 'local residents have demanded safer streets', 'local' and 'safer' are adjectives describing their respective nouns.

Adverb Most adverbs end in *-ly*. There are two types. First, verb-phrase adverbs that say how the action in the verb takes place, such as the words underlined here:

The politician <u>quietly</u> but <u>firmly</u> argued for more investment in the railways.

Second, sentence adverbs that show the speaker's attitude to what's being said:

Understandably, the miners demanded better pay and conditions. The coal-owners, not surprisingly, refused.

Clause A group of words, often not a complete sentence, containing such things as a doer and verb. There are two clauses in: 'If there were no bad people, there would be no good lawyers.'

Contraction A word with one or more letters missing and replaced by an apostrophe, for example: *don't* (do not), *won't* (will not), *haven't* (have not), *can't* (cannot), *I'd* (I would or I should). In informal business writing, contractions can add a little conversational warmth. The more unusual ones, like *you'll* (you will), *there'll* (there will), *there're* (there are) and *you'd* (you would) look odd and are best avoided in formal writing.

Co-ordinator A word that links two or more words, sentences or clauses, for example: *but, when, and, yet, if, although*.

Doer The person doing the action in a sentence, also known as the agent. In 'The publisher sent the contract', 'publisher' is the doer.

Grammar The body of rules and conventions by which words are grouped in a way that is meaningful to other people. For example, the sentence 'Paris look a beautiful city' is ungrammatical – if standard English is the criterion – as it breaks the rule that a singular doer must govern a singular form of the verb. Non-standard English uses different grammar which can be just as effective in the right circumstances. Bob Marley, the reggae musician, sang of his ancestors: 'Pirates, yes they rob I/Sold I to the merchant ships/Minutes after they took I/From the bottomless pit.' And usage changes over time. Queen Victoria was quite happy to write 'The news from France are very bad' because in her day 'news' was an accepted plural.

Imperative The form of the verb that gives commands, for example: *go, eat, push, don't jump, let them go, let me see, don't be deceived.*

Infinitive The basic form of the verb, made up of *to* plus the verb word. Present tense infinitives include *to go, to eat, to dream.* Past tense infinitives include *to have gone, to have eaten, to have dreamt.* Passive voice infinitives include *to be eaten, to be attacked.*

Nominalization A noun phrase formed from a verb, for example *preparation* (from *to prepare*), *renewal* (from *to renew*).

Noun A word that signifies a person, thing, place, activity or quality, for example *axe, beacon, carrot, dam, eating, frill, gerbil, happiness.* Proper nouns are specific names of people, places and the like: *Christmas, Canberra, Canada, Caroline.*

Paragraph A sentence or group of sentences separated in some way from the rest of the text and dealing with a particular part of the topic being discussed.

Participle The present participle adds *-ing* to the verb, hence *going, finding, sleeping.* The past participle adds *-d* or *-ed* to most verbs, as in *worked, decided, starved.* See chapter 4 for the way other past participles are formed.

Plural More than one. 'Frogs' is the plural of 'frog', which is singular. Where optional plurals are available for words of Latin or Greek origin, favour the English. Hence *referendums* not *referenda, forums* not *fora, stadiums* not *stadia, formulas* not *formulae.*

Preposition A word that usually comes immediately before a noun or pronoun, such as *in, down, up, under, of, with, by, to, from, at.*

Pronoun A word that stands in place of nouns, for example *he, she, it, him, you, I, me, they, anyone.*

Sentence A statement, question, exclamation or command – usually starting with a capital letter and ending in a full stop – which is complete in itself as the expression of a thought.

Singular See Plural.

Tense This refers to when an action occurs:

- Present tense: *he goes; he is going; she survives; she is surviving.*
- Future tense: *he will go; she will survive.*
- Past tenses: *he went; he has gone; he had gone; she survived; she has survived; she had survived.*

Verb Verbs express an action (*eat, sleep*) or a mental state (*know, believe*), so they are often described as doing words. A more accurate description is 'time-action words' because the action takes place in past, present or future time. In active voice verbs, the doer normally comes in front of the verb: *Goats destroy vegetation; Pigs might fly.* In passive voice verbs, the doer normally comes after the verb: *Vegetation is destroyed by goats,* or is not present at all. See chapter 4 for more on this.

14 Planning effectively

Guideline: *Plan before you write.*

'I don't know how to get started' is a common complaint. In the 'weaknesses' section of her pre-attendance form, one of my writing course students declared: 'Before I start I often have to think for some time about what I am going to write.'

Thinking before writing is usually a strength, and many people find that the best way to start writing (or dictating) is not to write but to plan. The first stage of planning is to think out:

- who is going to read the document;
- what they will be expecting to get from it;
- in what circumstances they will be reading (in their leisure time; at work in a quarry or on a building site; underwater);
- what you're trying to achieve.

It is better to plan on paper or on a computer screen than in your head. This is because the brain is better at marshalling ideas when they are set down on the page. It's difficult to hold 20 points in your head while simultaneously assessing them for strength and relevance and grouping them into bundles of related points. For one thing, you're having to assess which of the two million billion possible ways of ranking the points is the best.

For a short document, keep the plan simple. Make a list or bubble diagram containing all the points you expect to make, in no particular order. (Opposite is a bubble diagram of my first thoughts for writing chapter 20.) Cross out the irrelevant points. Link up the rest into groups of related points. Rank them using one of the structures in chapter 15. Then you will have a good framework from which to write or dictate. Don't skimp this stage, but don't overdo it to the point of boredom.

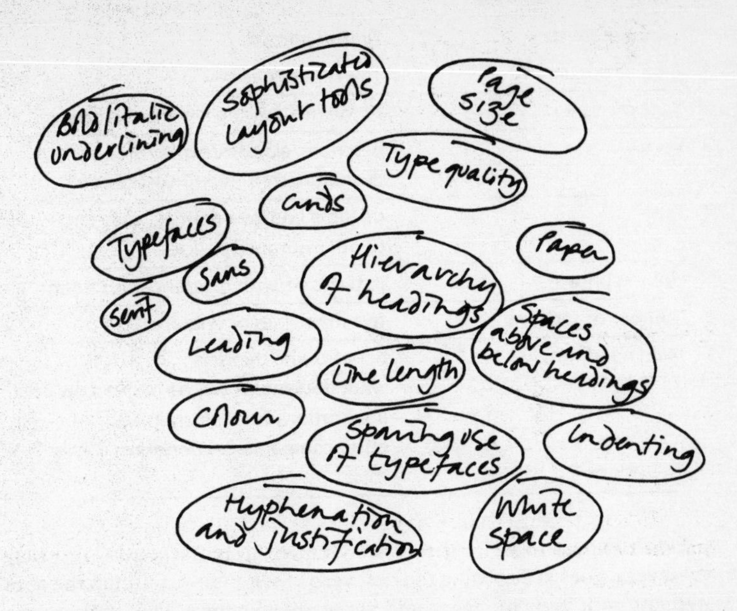

Longer documents benefit from fuller planning. If your managers ask you to write a report, discuss exactly what's wanted so that you are clear on the purpose and the amount of detail expected – otherwise you'll waste time on soul-destroying rewrites. If your managers don't really know what they want, you should investigate the topic thoroughly and return with a core statement (explained next) which sets out the purpose of the document as you see it.

Creating a core statement and horizontal plan

A core statement says what you will cover in the main section of the document – normally the discussion section. It helps you to focus on the task and the audience. Sharpening your focus in this way will save you researching topics that don't need to be covered. The core statement also builds your confidence as it provides your first glimpse of the finishing line.

The core statement is a rigidly constructed sentence in seven segments, as shown overleaf. The oblique strokes separate alternative words and phrases that might be used:

	Core segments	*Typical phrases*
1	Type of document	This report/paper
2	Your readers	to the head of the legal department
3	Verb	describes/assesses/explains/analyses/evaluates/considers/investigates
4	Topic	possible improvements to the clarity of our insurance policies
5	Linking phrase	in terms of/with reference to/under
6	Number of sections	five main lines of enquiry
7	Main headings	– benefits and dangers – what documents will be worked on? – implications for staff training – other companies' experience – costs

Don't expect to write the core statement in ten seconds. You may need several goes at choosing the best verb in segment 3. Discuss the core statement with your manager and agree any changes. This reduces the chance of creating a report that nobody wants.

The core statement underlies the next stage of planning – the horizontal plan. Regard this as a series of related boxes into which you slot all the points you expect to make under the headings set by the core statement. The main practical points are:

- Use the biggest sheet of paper around (prefer A3 to A4).
- Rotate the page so that the longer edge is nearer to you.
- Set down your headings along the top.
- Jot down all the points you can think of under that heading. Usually you will want to create a system of sub-headings within each section: boxes nested within boxes.

Let's assume the core statement said:

This report to the head of the legal department describes possible improvements to the clarity of our insurance policies under five main lines of enquiry: benefits and dangers; what documents will be worked on; implications for staff training; other companies' experience; costs.

A much-simplified (and tidied up!) horizontal plan for two sections of the report might look like this:

Benefits and dangers	Complications for staff training
Benefits	Direct sales staff
• Focus for an advertising + sales campaign	• No new training required – in effect same policies.
• Get in line with marketplace – customer satisfaction	• Explain purpose and discuss likely customer responses.
• Clear English Standard	• New selling points (clear, simple, easy to use).
• Compliance with EU directive	
Dangers	Head Office staff
• Legal safety – consult lawyers/counsel	• Briefing on new features for all in direct customer contact.
• Confusion among existing customers?	• Explain philosophy.

You can then assess the points for strength and relevance and, if necessary, group them under further sub-headings. Gaps in your knowledge may be revealed; you may need to fill them before proceeding. Next, put the headings and points in order, perhaps using one of the structures in chapter 15.

The writing itself can now begin. As you write you will probably change your mind about the order of points, and add to the plan. This is fine – the plan is meant to be a working document. One tip when you do start to write: there is no need to begin at the beginning. Provided they are self-contained, the bits you are confident about and know best can be written first. Then you will be less anxious about tackling the rest.

Alternative approaches to planning

Writers I have worked with seem to perform better when they plan. They also *believe* they write better and more quickly. But not everyone is happy to plan to the same extent: there are differing approaches, all with their followers. For convenience, four main labels can be used to represent these approaches: 'architects', 'watercolourists', 'oil painters' and 'bricklayers'. You may like to consider which categories you fit into and whether the alternatives might suit you better. Bear in mind, though, that most writers fit into two or more of the categories from time to time.

- **Architects** tend to take a three-stage approach to writing: planning, writing and revising. They do not try to perfect the writing as they create the first draft. They prefer to leave a draft and come back to it later, revising thoroughly on paper (not on screen).
- **Watercolourists** try to produce a complete version rapidly at the first attempt, with little revision. There may be little planning on paper but long mental incubation of the document. Some literary writers are said to favour this, believing that it keeps their writing spontaneous and true to feeling.
- **Oil painters** tend to jot down ideas and then organize and repeatedly rework them. On paper there is little planning. Rewriting seems to help oil painters to find out what they think, rather than helping them to focus on the readers' needs.
- **Bricklayers** tend to polish every sentence before going on to the next, in a careful process of building block by block. This means they revise very little once a first draft has been prepared. Bricklayers tend to plan and have a clear idea of what they want to say from the outset. They tend not to regard writing as an aid to thinking.

Given the importance of careful thought about the readers' needs when preparing plain English documents, the architectural approach, with its bias towards planning, seems to be the most suitable. Again, though, you should be ready to experiment, use different strategies for different tasks, and do whatever will get the writing job done to a satisfactory standard with the minimum of effort.

Strategic planning: learning from readers

A strategic plan may be necessary, especially when preparing a major document. The plan below relates to the preparation of a booklet for staff joining a company pension scheme, but many of its points are universal. The plan builds in plenty of reader involvement – essential unless secrecy, deadlines or costs prevent it.

Pre-production stage

- Decide purpose. Why is the booklet needed? Would something else do a better job?

- Determine the content. At this stage, just outline main points.
- Define audience. Consult staff to see what they expect from the booklet. Discuss the outline. If possible, get an idea of how they would like the information organized and even what they'd like it to say.
- Consider how the booklet fits into the system. How will it be distributed, used, stored? How will it be amended in future? What organizational politics will need to be dealt with and how?

Production stage

- Fill in the outline of the content. Decide what structure is most suitable.
- Write the draft. Include headings and sub-headings but don't worry about adding too many layout features at this stage.
- Evaluate the draft with staff – include some who haven't yet been consulted. Use a questionnaire to test comprehension. Iron out problems by redrafting or even restructuring.
- Prepare an appropriate layout. Use professional help if necessary. Evaluate the layout with staff.

Post-production stage

- Final evaluation: after distribution, use a survey and face-to-face discussions to gauge reactions. Did the booklet do its job?

15 Using reader-centred structure

Guideline: *Organize your material in a way that helps readers to grasp the important information early and to navigate through the document easily.*

You might be lucky. Your readers might regard your letter, memo or report as the highlight of their day. They might give it the time you know it deserves. They might sit back in a comfy armchair and browse through it contentedly for hours.

Or you might be like most other writers. Your document will be an interruption to your readers' busy day. They'll read it in haste, scribbling remarks on it and picking out important bits with a highlighter pen. And they'll be asking two questions with every sentence they read:

- So what?
- How does this affect me?

Answering these questions in plain words and short sentences won't always be enough. You will also need to organize the material so that readers can extract what they want in the shortest possible time. This will increase the likelihood of getting your ideas read – the first step in getting them accepted. And a crucial way of organizing material well is to put important news early. Then readers can see, right from the start, exactly what you're getting at.

This chapter shows ten ways ('models') of organizing your information. Don't feel that you have to force your writing to fit the models or, where headings are suggested, that you have to use the exact words. Quarry them for ideas that will suit your own way of doing things.

Model 1: Top-heavy triangle

Put your most important point first, follow it with the next most important, and so on, until your last paragraph includes relatively minor points.

You may need to do a bit of scene-setting to start with: use the first sentence or the heading for this:

	John
	Request for leave from 25 March–12 April
Scene-setting	Thanks for your e-mail message yesterday.
Big news	The managing director has agreed to your request to take special unpaid leave immediately after your visit to Athens. It is vital, though, that we have a brief written report of your key findings as soon as the business part of the trip is completed. Will you please fax this to us on 24 March.
Less important	As your flight was booked some time ago, please liaise with Michael South about rearranging the return leg.
Minor points	Hope this helps. We'll speak again after Tuesday's meeting.

Even unwelcome news should usually be delivered early. You may need to soften it, though:

	John
	Request for leave from 25 March–12 April
Scene-setting	Thanks for your e-mail message yesterday.
Big news	The managing director considered your request carefully but I'm sorry to tell you that he has had to say no. He feels that the launch of the new product at the end of March really demands your presence here.
Minor points	I realize you'll be disappointed but perhaps we can have a chat about things after next Tuesday's meeting.

I know of only one good reason for saving big news till the end – deliberate obfuscation, when delay will increase surprise, raise tension or leave the reader with a sense of threat.

The top-heavy triangle is useful when you're asking for something your readers may be reluctant to give. The letter below is trying to get local traders to fill in a questionnaire. But where does this vital information come?

Dear Mr Jones

Our Sea Fishery Group is conducting studies in order to aid the development of a local Fishery Forum. The idea of a Fishery Forum is to promote the fishing industry of the north west and to stress the importance of a regional approach as this could attract support from the European Union.

The study is designed to help evaluate the monetary impact that fish landings from the north west fishing fleet have on the region as a whole. Also the study will assess the feasibility of placing a logo on local fish products so that their origins can be instantly recognized.

To gather information for the study, a questionnaire has been designed. It is hoped that a high proportion of these forms will be completed and returned so that the final analysis will be as detailed as possible.

All answers to the questionnaires will be confidential and only the combined figures for the whole region would be published.

I would greatly appreciate it if you could complete the enclosed questionnaire by 15 June and return it to me. If you have any queries about the study or the questionnaire, please contact me on the above number. Thank you for your help.

The readers only learn the main action point in the fifth paragraph, and they might never get there. It could have been better to begin:

Dear Mr Jones

IMPROVING THE PROSPERITY OF THE NORTH WEST

I'm writing to ask for your help with a venture that will ultimately improve the prosperity of our region and of every trader who does business here.

Our Sea Fishery Group is conducting research into setting up a Fishery Forum that would promote our local fishing industry.

As part of the research, I need your help. Would you please be kind enough to complete the enclosed questionnaire and return it to me by 15 June? [etc]

With longer documents that have a number of headed sections, you could use the top-heavy triangle in each section.

Model 2: Problem–Cause–Solution

This is a simple model for short reports. First you state the problem . . .

> **You asked me to find out how a batch of chocolate came to be contaminated during the night shift on 25 April.**
>
> Then you state the cause . . .
>
> **What seems to have happened is that some inadequately treated reclaimed chocolate was added to the mix. This occurred because the reclaimed chocolate was mis-labelled and stored in the wrong part of the factory.**
>
> Then you say what should happen in the future . . .
>
> **The contaminated batch will have to be destroyed. The cost of the loss is about £2,800, taking everything into account. I have told supervisors to tighten their procedures for labelling and storage. They'll be reporting back to me next week.**

Model 3: Chronological order

This simply follows the time sequence of a series of actions. For more on this in its application to instructions, see chapter 18.

Model 4: Questions and answers

Questions and answers help to break up the information into small chunks. Here is an excellent example which also benefits from short paragraphs and everyday English:

> **How do I make a complaint about the council's services?**
> If you have a complaint, please take it up first with the department that deals with the subject of your complaint. Contact them and try to get the problem settled there and then. Phone numbers of departments are shown at the end of the leaflet.
>
> **What if I am not satisfied?**
> If you are not satisfied or your complaint is complicated, please set out all the facts clearly by making a formal complaint in writing. You can

> do this on the pull-out form opposite. If you need help to complete the form, please ask at any council office. We will acknowledge your complaint and keep you informed at every stage.
>
> **Where else can I get help?**
> If you are not satisfied with the way your complaint is dealt with, you could contact one of your ward councillors. A list is available at all of our reception desks and from Member Services. (Telephone: Cadeby 4000.)

Questions tend to provoke interest in readers by bringing them into the action. Questions are particularly useful if they use personal words. Compare:

Where can more information be obtained?

with

Where can I get more information?

Questions also convert dull, plodding label-headings into verb-rich information. Compare this heading:

Frequency of use by paramedics of aspirin in the treatment of heart attack victims

with its equivalent question:

How often do paramedics use aspirin to treat heart attack victims?

or, in the passive to put the stress on 'aspirin':

How often is aspirin used by paramedics to treat heart attack victims?

Model 5: S-C-R-A-P (Situation, Complication, Resolution, Action, Politeness)

A corny mnemonic, but useful. A letter using this structure might say:

	Dear Mr Soaring
	Order for 5 Sky-Fly hang gliders, £2,250 each
Situation	Thank you for your order dated 17 December; my apologies for the delay in responding.
Complication	The manufacturers have recently withdrawn the Sky-Fly and now offer a much improved

model, the Sky Jet. I enclose a leaflet which gives details of all its features.

Resolution	Although the price of the Sky Jet is £255 more than the old model, I can offer it to you at a special introductory price of £2,385 until 10 January next. We could let you have immediate delivery.
Action	If you would like to order the new model at the discount price, do please give me or my assistants a call.
Politeness	I look forward to hearing from you.

Model 6: S-O-A-P (Situation, Objective, Appraisal, Proposal)

Here's a much-simplified short report that uses SOAP:

FALL IN SALES OF RABBIT CHOCOLATE BARS

Situation
In the last three months, Rabbit has lost about 10 per cent of its share in the aphrodisiac chocolate bar market.

Objective
To regain and improve market share in the next six months.

Appraisal
The impact of our initial advertising and promotional burst has petered out. Research shows the recent price increase didn't affect sales very much but Rabbit has started losing ground to similar-priced competitors like the Hunky bar and the Rhino bar: people are seeing them as more sophisticated products.

Proposal
Target the higher social groups with a more sophisticated campaign based on Beatrix Potter stereotypes, including special-offer shooting weekends in Brer Rabbit country.

Model 7: PARbox memos

Some firms, infuriated at the time-wasting woolliness of inter-office memos, insist that all begin with three standard boxes for Purpose, Action and Response:

Purpose To explain developments in the Savill fraud case

Action requested Your help in unearthing missing documents

Response required By 10 January

Readers can then immediately see the key facts and fit the request into their schedule.

After the PARboxes come a main heading and text as normal.

Model 8: The 5 P's (Position, Problem, Possibilities, Proposal, Packaging)

Using these headings, a short and very simplified report for a sports club might say:

UPDATE ON OUR LEAGUE PLACING

1 Position

1.1 With only 10 games to go this season, the team is close to the relegation zone.

1.2 The coach's contract has two years to run at £70K a year. He is widely respected in the town and throughout the sport.

2 Problem

2.1 Recent results have been very bad – 9 defeats and 2 draws in the last 11 games. Attendances are down 10 per cent on last season's and our overdraft stands at £3m. The fans are restless and some have started barracking the coach.

3 Possibilities

3.1 We could buy new players before the transfer deadline – the coach's proposals are attached.

3.2 We could terminate the coach's contract immediately but this would be costly and very unpopular with most of the fans. It could also damage team morale.

3.3 We could leave things as they are.

4 Proposal

4.1 We should try to buy Player A to improve the defence. Apart from that, we should leave things as they are because:
 (a) there's little cash for new players or buying out the coach's contract;
 (b) two key players will soon return from injury;
 (c) the team has played better recently and the last two defeats were very narrow.

5 Packaging

5.1 It may be a cliché, but we need to express full confidence in the players and the coach and say that they would benefit from the continued support of the fans. We need to explain the club's financial position better. We need to reassure supporters that their views will continue to be considered.

In the report I have added a commonly used system of paragraph numbering. The aim is to help readers to navigate through the report and, if necessary, comment on it more easily. An alternative to a decimal system is simply to number paragraphs in 1-2-3 order. In either system, if you use a second tier of heading, you don't need to number it but you should distinguish it by layout, perhaps with bold type or underlining.

If possible, avoid creating a third level of decimals (3.1.2, etc). Some readers boggle at such complications.

Model 9: Correspondent's order

It's possible to respond to some letters or memos using the same order of points as your correspondent. Normally it's wise to say that this is what you're doing:

I'd like to respond to your points using the same order in which you made them.

Model 10: Full-dress report

A good arrangement for a detailed administrative or technical report is given below. Notice that the summary is placed early, though it will be written late in the process. A scientific report would also have sections for 'method' and 'results', which would follow the introduction, and perhaps an abstract (this is like a condensed summary that can be included in a database of abstracts).

Title

This says what the topic is, briefly but not too cryptically. If necessary, add a longer sub-title. Also add the author's name, the date and the distribution list.

Contents list

If included at all, a contents list deserves a page of its own.

Summary

This serves up the report's most interesting points in bite-size chunks. It should briefly set out the report's purpose, main findings, main conclusions and main recommendations. It is designed to give an accurate and rapid understanding of the main issues, enabling busy readers to ignore everything else if they wish. Therefore any main conclusion or recommendation that is qualified elsewhere in the report should also be qualified in the summary. A summary may contain sub-headings, which should differ from any other main headings in the report.

The summary should mainly be composed of informative not descriptive statements: 'Five staff should be transferred to Section X', not 'A recommendation is made about transferring staff'.

Remember that all your readers will study the summary, whereas perhaps only 30 per cent will read the full report – so allow proper time to write it well. In a short report (say, two or three pages), the summary can come after the conclusions and recommendations or be omitted altogether.

Introduction

This explains the purpose, background, how the work was done and what it cost. It may include brief, simple acknowledgments.

Discussion

(A heading that need not appear if more specific alternatives are available.) This sets out your main results, findings, arguments, options and ideas, preferably under headings and sub-headings so that readers can easily find their way around and skim from point to point. It is likely to be the longest section.

Conclusions

These are inferences you draw from the Discussion, not 'concluding remarks'. It may be more convenient to include them in the Discussion itself but putting them into a separate section may help the reader, especially in a long report.

Recommendations

These are the steps you think should be taken, based on the conclusions. Again, it may sometimes be convenient to include them in the Discussion. Recommendations should normally be written in the 'should' or 'should be' style.

Appendices

These provide a home for any fine detail that readers will not need in order to follow the Discussion.

If most readers are unlikely to want the full report, save trees by circulating only a summary with the title and distribution list attached. They can then call for the full report if they wish.

Finally...

Unless you are a novelist, poet or journalist, your writing will rarely be the highlight of a busy reader's day – but then it's not meant to be. Your main aim is to help readers to achieve their objectives which are, to put it bluntly, to Get In, Get On and Get Out. In other words, to access the information readily, make progress through it quickly and finish reading as soon as possible. Lucid language and structure will help them to do so, and that can only be good for your reputation as a clear thinker.

16 Using alternatives to words, words, words

Guideline: *Consider different ways of setting out your information.*

Everyone knows that the written word alone is not always the best way of communicating a message. Graphic devices such as tables, illustrations, pie charts, diagrams, maps, strip cartoons, mathematical formulas and photographs can all help. The difficulty is knowing when to use them and which ones will suit the information. You have to be flexible and experiment a bit to see what works. And, ideally, you should test your decisions with typical readers.

Using tables

An insurance company wants to tell a customer how much her policies are worth, so it uses continuous text:

> **1/ You now hold three equity plans. The 1992 plan is mature and is valued at £462.08. The 1993 plan will not become mature until 1 January 1995 and is currently valued at £645.43. The 1994 plan will mature on 1 January 1996 and is currently valued at £438.45.**

Sentences are short and simple enough for any investor who knows that 'mature' means that the policy is ripe for cashing in. (You noticed, though, that the last two sentences switched from negative to positive despite making similar points.) Using elaborate alternatives to text would be unjustifiable for such simple ideas. But a table would be feasible because the information in each sentence consists of a year, a value and a maturity date. Creating a table is fairly easy in most word-processing packages, and it might look like this:

2/ You now hold three equity plans, as follows:

Year of plan	*Current value*	*Maturity date*
1992	£462.08	Already mature
1993	£645.43	1 January 1995
1994	£438.45	1 January 1996

Or you could try it this way:

3/ You now hold three equity plans, as follows:

	1992 plan	*1993 plan*	*1994 plan*
Current value	£462.08	£645.43	£438.45
Maturity date	Already mature	1 Jan 1995	1 Jan 1996

Of the three possibilities, I prefer version 2, as it seems easier to compare the values and maturity dates in vertical columns than in continuous text or rows. The preferred table could be extended easily if, say, details of ten plans were being given. In continuous text this would be tedious, while you'd need a very wide piece of paper to accommodate layout 3 for ten tables.

Exploring other possibilities

The more complex the information, the more the possible choices of presentation multiply, and there is rarely one perfect solution. To show this, here are five ways of presenting the same handful of facts about what a local government department will charge for hiring a hall.

1 *Continuous text: official language version*

This version is the real one being used by the department. Though far from total gobbledygook, it doesn't seem clear at first reading:

Hiring fees of £100.00 or less must be paid in full at the time of booking. In the case of all hiring fees in excess of £100.00 the hirer agrees to pay £100.00 deposit at once on all bookings valued up to £200.00 or 50 per cent of the total hire fee as a deposit on bookings on or over £200.00 in value. The balance of the cost of each hiring must be paid at least 21 days before the date of the event.

I asked several writers to come up with alternatives. Shown below are the four main lines that emerged. In one or two, the tone of voice has changed with the use of 'please' instead of 'must'. (In a contract, 'must' would be necessary.) Otherwise, the content is the same. At the end of all the examples, I'll give the verdict of our focus group. Which do you think is best?

2 *Continuous text: plain English version*

This version paragraphs the information:

If a hiring fee is less than £100 please pay it in full when you book.

If the hiring fee is from £101 to £199, please pay £100 deposit when you book, then pay the rest at least 21 days before the event.

If the hiring fee is £200 or over, please pay half the hiring fee as a deposit when you book, then pay the rest at least 21 days before the event.

Because the sentence patterns are similar (all beginning with 'if' statements), readers can more easily pick out the details that are relevant to them.

3 List

The plain English version could become a list:

If the hiring fee is:

- **£100 or less, please pay it in full when you book.**
- **£101 to £199, please pay £100 deposit when you book, then pay the rest at least 21 days before the event.**
- **£200 or over, please pay half the hiring fee as a deposit when you book, then pay the rest at least 21 days before the event.**

4 Questions and answers

A question-and-answer set-up would also be feasible:

Is the hiring fee £100 or less?
Please pay it in full when you book.

Is the hiring fee £101 to £199?
Please pay £100 deposit when you book, then pay the rest at least 21 days before the event.

Is the hiring fee £200 or over?
Please pay half the hiring fee as a deposit when you book, then pay the rest at least 21 days before the event.

5 Table

The final option is a table (given to the focus group in bigger type than shown here):

Hiring fee	*Deposit*	*Balance to be paid*
£100 or less	None – pay fee in full on booking	
£101 to £199	£100 deposit, pay on booking	At least 21 days before event
£200 or over	Half hiring fee, pay on booking	At least 21 days before event

The focus group's verdict

Apart from the first, I would class any of the different presentations as plain English. On a points basis, the focus group narrowly preferred version 4, giving it an average of 18 out of 20 for clarity. The differences were small, though: version 5 scored 17; version 3 scored 16; and version 2 scored 15.

In terms of first preferences, version 5 was the clear leader: 11 people out of 35 preferred it, while 8 preferred version 4.

Version 1 was a distant fifth, scoring 6 points out of 20 for clarity and receiving no first preferences.

These results suggest that the focus group preferred plain English in whatever layout it appeared, but that they regarded layout as an important factor in their ultimate preference between the plain English versions.

In practice, the best presentation would depend on the purpose of the document and the readers' purpose in reading it. A question-and-answer layout could be seen as friendlier than a table, but such considerations might be more relevant in a sales leaflet than a legal contract. Readers wanting to get the answers to specific questions might perform better with a table, while those wanting to remember what they had read might perform better with questions and answers.

There are no easy solutions: writers need to know what presentation methods are available, be flexible in using them, and try to find out their readers' reactions to them.

Algorithms

An algorithm may be a good solution if the information aims to help people make a decision from a complicated range of possibilities, or to explain a complicated rule or procedure. Algorithms tend to occupy more space than text and may take a long time to write and lay out, but they can save the readers a great deal of effort because a thousand words of text may be convertible into a page of algorithm. The main danger is that readers may be so put off by the unusual format that they will reject it immediately.

Case history: A charity wanted to gather donations in the most tax-efficient way. It explained several methods, 'Gift Aid', 'Deed of

Covenant', 'Payroll Giving' and 'Deposited Covenant'. For simplicity, I haven't included the explanations here. The fundraising leaflet said this:

If you pay UK income tax and:

- **you want to make a one-off donation of £250 or more, please use Gift Aid;**

or • **you want to make a one-off donation of less than £250, please use a Deposited Covenant;**

or • **you want to give regularly and your employer has a payroll giving scheme, please use the scheme;**

or • **you want to give regularly and your employer does not have a payroll giving scheme, please use a Deed of Covenant;**

If you do not pay UK income tax, we will not be able to reclaim tax on your donation but please make a donation anyway.

This seemed easy enough to follow and the charity was satisfied with the level of donations. The next year, however, it decided to put the information into the algorithm shown in the box opposite. Though other factors may have influenced the result, the charity found that its donations increased by a fifth.

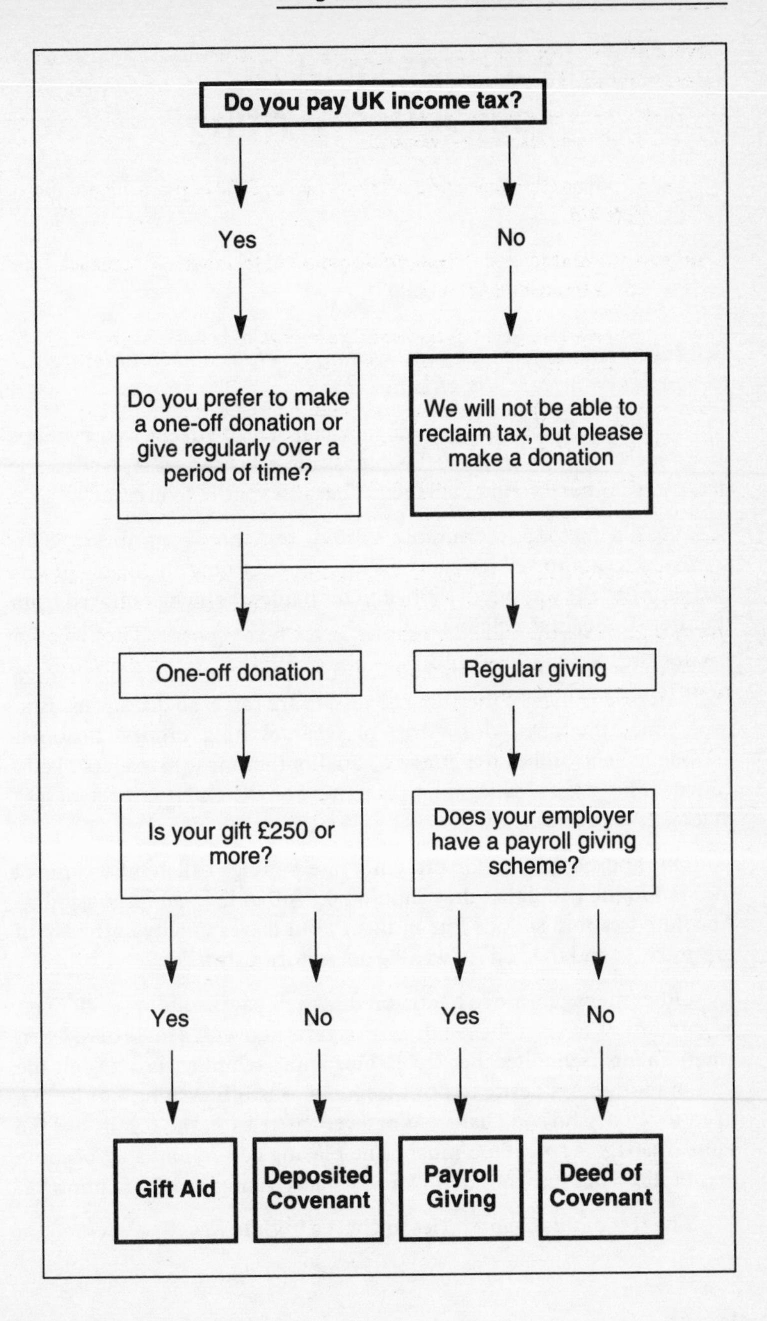
Do you pay UK income tax?
Yes
No
Do you prefer to make a one-off donation or give regularly over a period of time?
We will not be able to reclaim tax, but please make a donation
One-off donation
Regular giving
Is your gift £250 or more?
Does your employer have a payroll giving scheme?
Yes
No
Yes
No
Gift Aid
Deposited Covenant
Payroll Giving
Deed of Covenant

17 Management of colleagues' writing

Guideline: *Manage colleagues' writing carefully and considerately to boost their morale and effectiveness.*

The managing director of a prominent chain of high-street jewellery shops gave a speech in which he said: 'People ask me how I can sell such attractive decanters so cheaply. And I tell them: "because they're total crap".'

A plain message, admirably candid, but the wrong message for someone trying to sell jewellery. Within a short time the firm's profits had slumped, he was out of a job, and his name was being removed from the shop-fronts in the chain.

Writers don't usually make such spectacular mistakes in letters to customers, and even if they do, the results are rarely so disastrous. But, over time, the drip, drip, drip of weak writing erodes customer confidence and sullies the image of quality that most managers like to convey. This makes management of writing an extremely important task that takes care and effort.

One approach is 'hands-off' – not intervening at all in the writing of the staff in the belief that they should take full individual responsibility. Absolute freedom sounds fine in theory but leaves quality in the lap of the gods . . . and the gods of writing are notoriously fickle.

Other managers are so hands-on that their paw prints are evident on every word. Whatever their staff write is returned with a mass of red-pen amendments, schoolteacher style. This approach may spring from the best of motives – a desire to produce lucid, unstuffy writing – or from a need to show who's in charge. Whatever the reason, the results are the same: managers spend too much time playing editor, and staff become demoralized because they don't know what is wanted or how to improve.

The rest of this chapter tries to chart a middle way.

Working with the team

For the sake of example, let's assume you are managing a team which answers complaints from customers. Though team members deal with many complaints by phone, they also write letters. Here are a dozen things you could do.

1. Ask the team how they recognize a good complaint response when they see it. Usually they will agree with you (and me) that the letter should be relevant, accurate, well organized, clear and concise. It should seek to rebuild customer relations and even strengthen them beyond the pre-complaint level.

2. Ask the team what you can do to help them achieve these things. You might be told a few home truths about how you should supervise their writing but you may also get requests for extra training, style guides, other books on better English, dictionaries and electronic style checkers.

3. Show one of your own first draft letters and ask the team to compare it with the final draft you sent. Ask for ideas on how the final draft could have been improved further. Stress that nearly all writing can be improved by the injection of another person's ideas.

4. Show some of the team's letters and ask them to pick out strengths and weaknesses. If there are problems of spelling, punctuation, sense and tone of voice, team members will realize for themselves that they will have to improve. Foster an atmosphere in which writing can be discussed in an unthreatening way.

5. Discuss the process of writing, particularly the need for planning by writers who use dictation machines or who type or speak directly onto word-processors. Stress that planning will usually save time and raise quality.

6. When you ask a writer to respond to a long or difficult complaint, ask him or her to come back with a detailed plan of the key points. Discuss and agree it before the writer produces a full draft. This will make purpose and content clear at every step and ensure that the writer follows a clear track from the outset.

7. Explain what standards you will use to supervise writing. You will be checking for:

    - *Purpose:* Is the reason for the letter clear to the reader?
    - *Content:* Does the letter contain the right messages and

deal with any relevant points the complainant made? Is it accurate, complete and relevant? Is it defensible in the light of your organization's policies or the laws or regulations that govern you? If it is based on a standard letter, has it been tailored appropriately to the individual's circumstances? Does it say what, if anything, the reader should do next?

- *Structure:* Is the material organized in a logical sequence? Does it place the big news early? In a long letter, are headings used to help the reader navigate? If there is an apology, is it positioned early enough? Refer to the various models in chapter 15.
- *Style:* Are the points explained clearly, in sentences of reasonable length with short paragraphs, good punctuation and without officialese and business clichés? Is it as warm and friendly as the topic allows?

8 Say that you will try to intervene as little as possible in the writing and that your interventions will seek to make genuine improvements without nitpicking. Your overriding question will be 'Is this letter fit to go?', not 'Is it perfect?' or 'Is it how I would have written it myself?'.

9 Promise that you will explain when staff ask why you have made substantial alterations to their drafts.

10 Make available the tools and information that will help staff to do a good job.

- Buy a good dictionary in book or electronic form.
- Copy and circulate press articles about words.
- Consider asking known good writers in your department to act as coaches and advisers.
- Make your copy of this book available to all. (Or buy several, say the publishers.)
- Add an effective style-checking program to your word-processing software. All style- and grammar-checkers have limitations because they cannot assess the sense, logic or structure of a document, but there are a few good ones.
- For six months or so, circulate letters from each month's output that you think are particularly good. Discuss them at your team meetings.

11 Trust your writers. With proper guidance, they should be able to produce good work without constant supervision. You will still keep a general eye on their writing by seeing copies of letters from time to time. Apart from that, let them get on with it.

12 Do a survey of recent complainants, including questions about whether they thought the response letters were clear and in the right tone of voice.

Making the right interventions

Let's consider two examples of writing that should have been substantially improved before they went out.

Case history 1. During a nationwide health scare about fresh eggs, a well-known firm of egg distributors (I'll call them Cockerel Brothers) asked hundreds of shop-keepers to reassure customers by displaying this notice:

> **Dear Retailer**
>
> **As a valued customer, for Cockerel Brothers. We would like to assure you of our continual monitering for anything which could affect the quality of our fresh eggs.**
>
> **Cockerels have just undertook a survey of all our feed mills which involved testing finished feeds and samples (scrappings) from the mills, and having them tested by the Public Health Laboratory. This has resulted in a programme of mill monitering to prevent any bacterial build-up in our mills.**
>
> **We do not use any raw poultry offal meat, and dried poultry manure in our mills.**
>
> **We are taking these precautions to endevour our feed is as free from harmful bacteria, as is humanly possible, and manufactured from only wholesome raw materials.**

Reassuring, it certainly isn't. Because few people expect hens to be fed with manure and offal, raising the possibility does not inspire confidence. And the errors speak of a certain carelessness – could their egg production be equally careless:

- The first sentence isn't complete and suggests that the writer is the customer, which would be nonsense.

- 'Undertook' should be 'undertaken', in standard English.
- 'Monitering' should be 'monitoring'.
- 'Endevour' should be 'endeavour'.

No supervisor of writing should have let the notice go out untouched. There should have been a friendly, tactful conversation in which improvements were discussed with the writer so that he or she could rework the text, perhaps trying to emphasize the wholesomeness of the product. A rival firm in similar circumstances decided that saying less was a far better way of saying more. Its notice simply said:

EDENDALE EGGS

Healthy Eggs from Healthy Hens.

Case history 2. A housing association is writing to several hundred tenants to explain why modernization work has been delayed. It's an important letter:

BATHROOM MODERNIZATION

First of all apologies for the delay in the start of the bathroom refurbishment programme which was due to start at the end of last year but because the costs came in over budget, further negotiations had to be entered into with the contractors and the extent of the work reconsidered.

Although all the work to the bathroom will be undertaken as previously agreed, we will not be undertaking any work to ground-floor WCs which you may or may not have.

Find attached a draft programme for the anticipated commencement date on your property and we anticipate that the work will take three or four days to complete. Your next contact will be by the contractor, GH Construction, who will contact you individually about a week prior to the start at your house.

If you anticipate any problems with access arrangements or require any further information, please do not hesitate to call Jane Teal on Tameside 000 99.

The first question a manager of writing needs to ask is 'Is this fit to go?'. My answer would be no. The writer has arranged the points logically and there are plenty of personal words like 'you' and 'we', but officialese, jargon, passives and long sentences are overused.

Great improvement is possible by following good writing tactics. Again you would need to discuss the letter, praise its good points but perhaps show a reworking of the first paragraph as an example of what to do with the rest. It would be useless – and demoralizing to the writer – to redraft the whole letter, though you could provide guidance notes in the margin. Helpful coaching, pleasantly given and co-operatively received, should enable the writer to produce a letter that you would be able to pass as 'fit to go'.

Taking one paragraph at a time, I have underlined possible insertions and shown deletions for the whole letter:

BATHROOM MODERNIZATION

~~First of all apologies~~ I apologize for the delay in ~~the start of~~ the bathroom ~~refurbishment~~ modernization programme which was due to start at the end of last year. ~~but~~ The delay has occurred because the costs ~~came in over budget,~~ were higher than we had budgeted for. ~~further negotiations had to be entered into and~~ We had to negotiate again with the contractors and reconsider how much work we could afford. ~~the extent of the work reconsidered.~~

This splits the long sentence into three and puts the verbs in the active voice.

~~Although all~~ All the work to ~~the~~ your bathroom will be ~~undertaken~~ done as previously agreed~~, we~~. But unfortunately the extra costs mean that if you have a ground-floor toilet, we will not be ~~undertaking~~ able to do any work to ~~ground-floor WCs which you may or may not have~~ it.

I have left the first verb in the passive because it will put the stress on 'all the work to your bathroom'. The other changes split the sentence in two and remove the burbling about 'may or may not have'. (It would, of course, be preferable to send one letter to tenants with ground-floor toilets, and a slightly different letter to the rest.)

~~Find attached~~ I attach a ~~draft~~ programme ~~for~~ which shows the ~~anticipated commencement~~ likely starting date for work on your property. ~~and we anticipate that the work~~ We expect the work will take three or four days to complete.

This splits another long sentence and uses plain English.

~~Your next contact will be by~~ You will hear next from the contractor,

GH Construction, who will contact you ~~individually~~ about a week ~~prior to the start~~ before the work at your house begins.

This strengthens the first verb and adds another in place of the noun 'start'.

If you ~~anticipate~~ think the contractor will have any problems with access ~~arrangements~~ to your house, or if you need ~~require~~ any ~~further~~ more information, please ~~not hesitate to~~ call Jane Teal on Tameside 000 99.

This costs a few more words in the cause of certainty but also gets rid of waffle. In the final, complete version below, the last sentence has been turned round so that the shorter 'please' clause comes first:

BATHROOM MODERNIZATION

I apologize for the delay in the bathroom modernization programme which was due to begin at the end of last year. The delay has occurred because the costs were higher than we had budgeted for. We have had to negotiate again with the contractors and reconsider the extent of the work.

All the work to your bathroom will be done as previously agreed. But unfortunately the extra costs mean that if you have a ground-floor toilet, we will not be able to do any work to it.

I attach a programme which shows the likely starting date for work on your property. We expect the work will take three or four days to complete.

You will hear next from the contractor, GH Construction, who will contact you about a week before work at your house begins.

Please call Jane Teal on Tameside 99099 if you think the contractor will have any problems with access to your house, or if you need any more information.

The focus group were asked whether they perceived any difference in clarity between this last version and the original.

The average clarity rating was 14 points out of 20 for the original, 17 for the revision. Twenty-five out of 35 people expressed a preference for the revision. The fact that 8 people preferred the original (with two scoring them equal) and that the scores were so close, is a salutary reminder that no-one can please all the people all the time and that audience reaction is rarely uniform. It could even suggest that the letter was fit to send in its original state.

Finally . . .

Criticizing someone's writing can be unnerving for them. So, before you intervene, always consider how you would feel if you were on the receiving end.

It is fair for you to want a high standard of clarity and grammar from all who write for you. But remember that no two writers will ever use exactly the same words or tone of voice, so trying to impose uniformity will not only be demoralizing, but futile.

18 Writing better instructions

Guideline: *Devote special effort to producing lucid and well-organized instructions.*

Everyone has favourite examples of ambiguous instructions:

- On a bottle top: 'Pierce with pin, then push off.'
- On the door of a health centre: 'Family Planning – please use rear entrance.'
- On a hotel breakfast form: 'Please hang outside your room before retiring.'
- In an aircraft maintenance manual: 'Check undercarriage locking pin. If bent, replace.' (The operator examined the pin. It was indeed bent, so he carefully put it back. The aircraft crashed.)
- On a tin of chocolate pudding: 'Before opening, stand in boiling water for ten minutes.'

A lifetime's experience of abysmal product instructions leads many people to read them only when all else fails. Hence the industry slogan 'RTFM' – Read The Flaming Manual. When his car clock failed, a top executive of an American car-maker tried to reset it using the approved company handbook. He couldn't. Next day he began a project to rewrite the whole manual in language that a $260,000-a-year corporate captain could understand. Such searing personal experience is often the spur to writing better instructions. Unless some handbook or manual has made you want to disembowel every employee in the company concerned, you might never feel the necessary passion to write it better.

Bad instructions are bad for business. Customers think twice about buying from a company whose instructions have proved useless before. In some countries consumer protection regulations require that instructions and safety information accompanying a product are taken into account when deciding whether it is faulty. Manufacturers can be held liable for injury or damage caused by poor safety information. The selling of unsafe consumer goods can lead to criminal prosecution.

So, what goes wrong in the writing of instructions, and how can problems be put right? Usually one or more of the following six commandments is broken.

Commandment 1: Remember the readers

Usually readers haven't used the product before, that's why they're reading the instructions. But what else can you find out about them? Will they be technically knowledgeable or complete novices? Will they be children or adults? Under what conditions will they be using the instructions?

Learning about these things and making the right assumptions will affect the words you use. An experienced carpenter might readily understand the instruction 'screw pendant bolts into door plates using cheese-head screws', but most do-it-yourselfers would need an illustration of pendant bolts, door plates and cheese-head screws.

Commandment 2: Favour a simple style of language

This often means using the command form of the verb, known as the imperative:

Switch on the computer

instead of

The computer should be switched on.

By favouring the imperative you are putting the action early and helping to keep the message simple.

Let's say you are instructing office staff how to fill in this box on a computer screen:

New Cost Centre Code
New Cost Centre Name
Parent Cost Centre Code

Using the passive voice you could write:

The code of the new cost centre should be entered into the New Cost Centre Code field.

The name of the new cost centre (maximum 40 characters) should then be entered into the New Cost Centre Name field.

The code of the parent cost centre should then be entered into the Parent Cost Centre Code field.

When satisfied, the 'Do' key should be pressed to commit the new cost centre to the database.

Or you could write it much more crisply in the imperative:

Enter the code of the new cost centre into the New Cost Centre Code field.

Enter the name of the new cost centre (maximum 40 characters) into the New Cost Centre Name field.

Enter the code of the parent cost centre into the Parent Cost Centre Code field.

When satisfied, press the 'Do' key to commit the new cost centre to the database.

This is simpler because the readers know from the start of each sentence what action is required of them. Even simpler would be to assign numbers to the three fields:

New Cost Centre Code	[1]
New Cost Centre Name	[2]
Parent Cost Centre Code	[3]

and then write:

Enter the correct code in field 1
Enter the correct name (maximum 40 characters) in field 2

and so on.

Commandment 3: Split the information into chunks

Readers waste time and make mistakes if the information they need is buried in long paragraphs. In this example a local authority inspector is trying to persuade a restaurant owner, whose dirty kitchens she has been visiting, to do his washing-up hygienically. Notice that there are no numbered steps or short paragraphs. The information looks boring and complicated, and you probably don't feel like reading it. That's just what the restaurateur must have thought.

> **After the preliminary sorting of the utensils and scraping off of food residues into the refuse containers, the utensils should be washed in the first sink, piece by piece, in clean hot water at a temperature of about 60°C with a detergent added. This temperature is too hot for the hands and the operative should wear rubber gloves and use a dish mop. The water should be changed as often as it becomes dirty or greasy. After this, the utensils should be suitably arranged in the wire baskets available for immersion in the sterilising sink. The utensils should be placed so that no two pieces touch each other and that all the surfaces of every piece are exposed to the rinse water. The rinse will be ineffective if plates or saucers are piled on top of one another or if cutlery is merely heaped in the basket. The sterilising rinse in the second sink should be of clean hot water without added detergent or chemical and at a temperature of not less than 77°C and the utensils should remain in the water for a full two minutes. At this temperature care should of course be taken not to immerse the hands. The purpose is to raise the temperature of the utensils to that of the water so that they will air-dry almost instantly on removal. The temperature of the water should be maintained at about 77°C throughout and accordingly this sink should be fitted with a device to record the temperature of the water. When the two minutes are up, the basket should be removed from the sink and stood temporarily on a draining board and as soon as the utensils are dry and cool enough to be handled, they should be put in a clean place awaiting re-use.**

The words are reasonably plain, the punctuation is sound and none of the sentences is impossibly long. But the whole thing needs splitting into short paragraphs ('chunking') and transforming into the imperative. First we need an introductory sentence, perhaps this:

> **You should make sure that your staff follow these instructions**

then the points need redrafting in the imperative:

1. **Sort the utensils and scrape off waste food into the bins.**
2. **Wash the utensils in the first sink, piece by piece, in clean hot water at a temperature of about 60°C with a detergent added.**

This temperature is too hot for your hands, so wear rubber gloves and use a dish mop.

3 **Change the water as often as it becomes dirty or greasy.**

4 **Put the utensils in the wire baskets available for immersion in the sterilising sink. Arrange the utensils so that no two pieces touch each other and that all the surfaces of every piece are exposed to the rinse water. The rinse will not do its job if plates or saucers are piled on top of one another or if cutlery is merely heaped in the basket. The sterilising rinse should be of clean hot water without added detergent or chemical and at a temperature of at least 77°C.**

5 **Put the wire baskets containing the utensils in the water for a full two minutes. At this temperature, take care not to immerse your hands. The purpose is to raise the temperature of the utensils to that of the water, so that they will air-dry almost instantly on removal. Maintain the water temperature at about 77°C throughout. To check this, make sure the sink is fitted with a suitable thermometer.**

6 **When the two minutes are up, remove the basket from the sink and stand it temporarily on a draining board.**

7 **As soon as the utensils are dry and cool enough to be handled, put them in a clean place awaiting re-use.**

These improvements have arisen not from a total rewrite but from two simple structural tactics (chunking and numbering), and one simple style tactic, the use of imperatives.

But are the 'improvements' perceived as such? Yes, according to the focus group. All 35 members preferred the revision. It received an average clarity rating of 18 points out of 20, compared to 10 for the original.

Commandment 4: Use separate headed sections

Normally it is wise to split the instructions into separate sections whose headings identify the purpose of each action.

A common sequence of sections is:

- *Introductory explanation, overview or summary:* This tells readers the purpose of the activity, what it will achieve and how long it should take. Sometimes users with experience of similar tools or equipment will benefit from a quick-start procedure. In long instructions, a contents list will help.

- *Tools or materials required:* Giving this information saves readers from having to stop the job whenever a new tool is needed.
- *Definitions:* These explain any terms the readers may not understand. Definitions may also be needed of everyday words carrying special or limited meanings in the instructions.
- *Warnings:* If these come after the instructions, they are useless and could be dangerous. Warnings should be given twice: once in the introduction and again just before the instruction to which they relate. Make clear that the warnings must be followed and are not just recommended practice. Could the product be modified to eliminate the hazard?
- *Main text,* split into headed sections.

The extract from a factory notice at the end of the chapter shows one way of splitting up text by using headings. Headings should be 'predictive' – that is, they should tell the reader what's coming up in the paragraph – and, if possible, stimulating. Label headings, like 'Spillages' or 'Storage' are weaker than 'What to do about spillages' or 'How much should be stored?'.

Commandment 5: Use appropriate illustrations, labelling and captioning them well

If a message can be simply conveyed in words, there is no need for illustrations (and vice versa). Words are particularly good for getting abstract ideas across, dealing with fine differences in meaning like 'could be' and 'might be', and for referring to things the reader has already learned about. Illustrations are good at showing what things look like and their relative size. This can save words and illuminate the words that remain.

Whether you are commissioning illustrations from graphics professionals or drawing them yourself, you may like to consider such things as:

- exploded-view format (useful for self-assembly products as they show how the item is put together);
- illustrations showing enlargements of particular parts so that readers can focus on the relevant point easily;
- the position of the illustration relative to the text – ideally readers should be able to refer to it as they read;

- presenting an object at an angle of view which clearly shows the parts concerned or the action to be taken;
- keeping objects to scale;
- ensuring that any typesetting included in the illustration will be legible if scaled down to fit the document.

These instructions for maintaining a pressure cooker lid integrate text and pictures in a straightforward way.

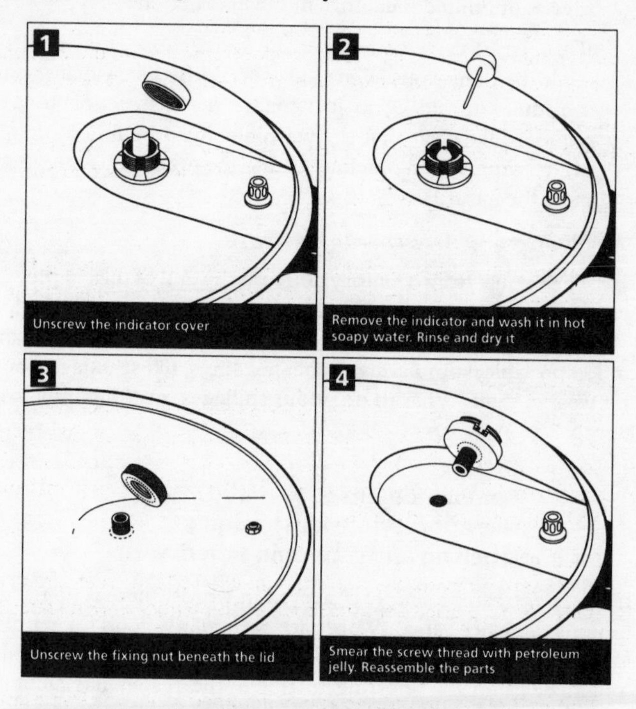

Commandment 6: Test with a panel of typical users

This is the most important commandment, because users' performance is the key. Give the draft instructions, and any product associated with them, to a focus group of typical readers – preferably not your own colleagues who will know too much about the product. Watch them trying to use the instructions. Don't intervene unless asked or unless there is danger. Observe any false moves they make. Discuss with them how they got on. Ask them about any misinterpretations. Redraft the instructions in the light of what you find. Test again if possible.

Code For Use And Maintenance Of Dominion Ink Jet Printers

Type of substances
Dominion inks use the solvent base methyl ethyl ketone, 'MEK'. MEK-based inks and solvents are all highly flammable and pose a serious fire and explosion risk if not handled properly.

Is it dangerous to breathe near the vapour?
In normal use there is no health hazard from inhaling the vapour. If spills are being cleaned up, wear respiratory protective equipment.

Avoiding skin contact
Dominion inks and solvents are not very poisonous but you should still avoid skin contact. If there is a risk of spills or when cleaning up spillages, wear solvent-resistant gloves and goggles.

How much should be stored?
Store as little of these substances as possible in the workroom. Never store more than 50 litres (including waste liquids). Storage should always be in metal cabinets purpose-made for highly flammable liquids.

Clearing up spills
Mop up small spills with industrial wipes or other dry material. After use, put the wipes in a purpose-made fire-resistant lidded bin. Absorb larger spills using a non-combustible material such as sand or Nilflam absorbent. Departments should ensure they have enough materials and equipment to deal with spills.

Dried inks
Deposits of dried ink are highly flammable, so take great care to avoid spills. Clean up dried deposits regularly using non-sparking scrapers. Dispose of the deposits safely.

Use of printers in safe areas
Printers should only be used and maintained in an area free from risk of ignition of flammable vapours. Keep the area well ventilated, particularly at low level. Eliminate sources of ignition such as naked flames and hot elements. Prohibit smoking. Use reduced-sparking tools where possible.

Fire
Suitable extinguishing agents are carbon dioxide, dry powder, alcohol-resistant foam and Halon. Large fires should only be tackled by trained firefighters. If containers get heated, isolate any nearby electrical supply. Then cool the containers by spraying them with water from a safe distance.

19 Lucid legal language

Guideline: *Apply plain English techniques to legal documents such as insurance policies, car-hire agreements, laws and wills.*

The way many lawyers write is disappointing to their friends and obnoxious to their clients. 'Excrementitious garbage' was how Jeremy Bentham described legal English in the nineteenth century. While every profession tends to cloak its mysteries in unusual language, lawyers have done so more thoroughly than most. But legalese is not sacred. Many lawyers despise it, realize it brings them into disrepute, and are working to remove it. In most English-speaking countries, groups of lawyers campaign for plain language. The UK group, 'Clarity', is chaired by a lawyer who pleads:

> **Why do lawyers write so that no-one can understand them? They say it is because they need to be precise, and that their language has been honed by centuries of litigation. But this is baloney. The real reason is that, although they are paid for their skill with words, most lawyers are dull and clumsy writers who have not broken the bad habits they learnt as students.**

No organization that wants to win the confidence of customers should give them incomprehensible legal agreements to sign. Now that there are many examples of clear, well-organized insurance policies, credit card conditions and tenancy agreements, it is simply uncompetitive to be locked into legalese.

The myth is slowly crumbling that obscurity is necessary or part of the price customers are willing to pay. Indeed, with some lawyers charging more for an hour's work than their customers earn in a week, it seems only justice that what they write should be understandable.

Traditional legal language creates imprecision and uncertainty. For example, this extract from a local law is meant to regulate child employment. It is framed in the traditional way with plenty of negatives and exceptions yoked together in each sentence:

No child under the age of 13 years shall be employed, except a child who has attained the age of 10 years may be employed by his parents in light agricultural or horticultural work on the parent's own premises.

No child shall be employed on a Sunday except in the delivery of milk or newspapers for more than two hours between the hours of 7–11am or in the milking and feeding of livestock for not more than two hours between the hours of 7am–7pm.

Now, simply ask yourself the sort of questions that, say, a newsagent might ask:

- May I lawfully employ a 14-year-old from 8am to 11am on a Sunday to deliver newspapers?
- May I employ such a child for only one hour during that time on a Sunday?

Straw polls among lawyers and local government officials reveal that these questions can be answered in so many different ways that the only certainty is uncertainty. Legal interpretation becomes as precise as astrology.

What can be done?

All the techniques described so far in the book can be used to good effect in legal documents. On their own, the techniques will raise the standard of legal writing, though they need to be harnessed to legal knowledge and a genuine care for precision and accuracy.

In the introductory chapter I said that readers needed to be co-operative and well motivated if they were to understand a text in the sense it was meant to be understood. With legal documents, the position is undoubtedly different: some readers will deliberately try to misinterpret a text in their favour. Certainty of meaning is therefore more important. But plain English, cautiously and appropriately used, can make an important contribution to certainty.

It is true that lawyers cannot blithely replace technical legal terms like 'negligence', 'indemnify' and 'estoppel' with one-word equivalents, though they can often provide separate explanations or glossaries. Yet in most legal documents only a few words are genuinely technical. The rest are plain words with ordinary meanings, or legal-flavouring words that smell of the law but can be replaced by plain words or struck out altogether.

This spells death, or at best very limited life, for the three ugly brothers 'hereof', 'whereof' and 'thereof', their kissing cousins 'herein', 'hereinafter', and 'hereinbefore' and their wicked uncles 'hereby', 'thereby' and 'whereby'. Such legal flavouring has virtually gone from modern UK laws, which proves how redundant it is elsewhere.

Even if you are not a lawyer, you can play a part in counteracting legalese by questioning its use wherever it appears. You may, for example, be asked to comment on documents written by your own organization's lawyers. You should be prepared to comment not just on the content but on its style and structure. The lawyers may not thank you for invading their territory, but you have something useful to offer – the ability to see the document as non-lawyers will see it.

The clarification of legal documents would make a book of its own, so I want to examine just three proven techniques:

- Replace or cut out legal flavouring.
- Chop up snakes.
- Put people into the writing.

You will probably have seen worse examples of legal writing than those that follow, but I have chosen them because they are understandable to the general reader while still displaying all the faults that need to be put right.

Replace or cut out legal flavouring

Case study 1. In this extract from a car-hire agreement, there is legal flavouring and pomposity (underlined):

> **In the event of car breakdown the owners must at their own expense effect the collection of the car and perform repairs thereto.**

The sentence is also ambiguous. I have heard lawyers argue that only the collection would be at the owners' expense, not the repair. But let's assume that both collection and repair are free to the hirer. We can then replace the legalese with plain words. 'In the event of' becomes 'if'; 'thereto' becomes 'to it' or 'to the car'; and 'must effect the collection' becomes 'must collect'. The sentence becomes:

> **If the car breaks down the owners must collect and repair it, both at their own expense**
>
> or
>
> **If the car breaks down, the owners must – at their own expense – collect and repair it.**

Two points on this:

- The ambiguity of the original has disappeared.
- The revision is shorter. Lawyers used to be paid by the word but now brevity costs them nothing.

Case study 2. A local government department is asking people who apply for a replacement car parking pass to sign a form. The legal flavouring is highlighted:

I hereby declare that the information given above is correct to the best of my knowledge and that I have conducted a thorough search for the said pass and honestly believe that the same cannot be found by me. I agree that in the event of the said pass being found by me I will forthwith return the same to the City Engineer.

All the legal flavouring can safely be removed:

- 'Hereby' just means 'by this writing'.
- 'The said pass' is no more specific than 'the pass' since no other pass is in question. If there were several passes, you would call them 'pass 1', 'pass 2' etc.
- 'The same' is legalese for 'it' or 'them'.
- 'Forthwith' has been given many different interpretations in courts of law – 'immediately' is plainer and would do just as well here.

The legal flavouring doesn't make the form impossible to understand but produces a fusty legal smell whose only purpose is to impress the reader with the solemnity of the procedure. But is it any less solemn in plain English:

I declare that:

- **the information given above is correct to the best of my knowledge;**
- **I have conducted a thorough search for the pass and cannot find it.**

I agree that if I find the pass, I will immediately return it to the City Engineer.

Case study 3. Pruning shears are the most appropriate tool when examining this example, in which a city requires a householder to cut back his hedge. Again, the legal flavouring and pomposity are underlined:

Whereas a hedge situate at Dean Road, Moreton belonging to you overhangs the highway known as Dean Road, Moreton aforesaid so as to endanger or obstruct the passage of pedestrians.

Now therefore the Council in pursuance of section 134 of the Highways Act 1959 hereby require you as the owner of the said hedge within fourteen days from the date of service of this notice so to lop or cut the said hedge as to remove the cause of danger or obstruction.

If you fail to comply with this notice the Council may carry out the work required by this notice and may recover from you the expenses reasonably incurred by them in so doing.

If you are aggrieved by the requirement of this notice you may appeal to the magistrates' court holden at Moreton aforesaid within fourteen days from the date of service of this notice on you.

The legal flavouring falls prey to plain words:

- 'Whereas' is fairly common and well understood, but here its use is pointless and creates an incomplete sentence. It can be struck out.
- 'Situate' means 'situated' or 'at'.
- 'Dean Road, Moreton aforesaid' just means 'Dean Road, Moreton'. If there were two Moretons, the correct one would need to be specified by adding the county – 'aforesaid' doesn't aid precision.
- 'Now therefore' is redundant.
- 'In pursuance of' can be replaced by 'under' or 'for the purpose of' or 'using'.
- 'Hereby' is redundant.
- 'So to lop or cut the said hedge' just means 'to cut the hedge' or 'to trim the hedge'. The New Shorter Oxford dictionary says 'lop' means to 'cut off the branches, twigs etc from a tree', so there is no need for both 'cut' and 'lop'.
- 'Fail to comply with' is a pomposity for 'disobey'.
- 'Incurred' is an unusual though very useful word. For a mass audience, the sentence might need rewriting to remove it.
- 'Holden' is a medieval remnant, meaning 'held'.

With dross swept away and new words underlined, the notice could read:

> ~~Whereas~~ You are the owner of a hedge ~~situate~~ at Dean Road, Moreton ~~belonging to you~~ which overhangs the highway ~~known as Dean Road, Moreton aforesaid so as to~~, endangering or obstructing ~~the passage of~~ pedestrians.
>
> ~~Now therefore the~~ The Council ~~in pursuance of,~~ under section 134 of the Highways Act 1959, ~~hereby~~ require you ~~as the owner of the said hedge~~ within fourteen days from the date of service of this notice ~~so~~ to ~~lop or~~ cut the ~~said~~ hedge, removing ~~as to remove~~ the cause of danger or obstruction.
>
> If you ~~fail to comply with~~ disobey this notice the Council may ~~carry out the work required by this notice~~ choose to cut the hedge and ~~may~~ recover from you ~~the expenses reasonably incurred by them in so~~ its reasonable expenses in doing so.
>
> If you are aggrieved by ~~the requirement of~~ this notice you may appeal to the magistrates' court ~~holden~~ at Moreton ~~aforesaid~~ within fourteen days from the date of service of this notice on you.

That draft would be fit to show to a lawyer to check for legal accuracy.

The focus group agreed that the revision was much clearer. The 34 who responded gave it an average clarity rating of 18 points out of 20. The original notice averaged only 8. Three police officers in the group, accustomed and perhaps a little loyal to the peculiarities of law language, each gave the original 15 points.

Chop up snakes

Long sentences are another blight in legal documents. Sentences in bank overdraft agreements have been known to stretch to 900 words, monstrosities that show contempt for the readers as well as befuddling them.

This 83-word sentence from an office equipment lease between Bigg, the owner, and Tiny, the company renting the equipment, explains what happens if the equipment breaks down:

> If the equipment shall go out of order, Tiny shall at its own expense have the equipment repaired by the person, firm or body corporate

designated by Bigg and in the event of Tiny failing so to do then Bigg shall be entitled to take possession of the equipment and have it repaired at the cost of Tiny and during such possession and repair, the lease charges shall nevertheless accrue and be payable by Tiny to Bigg.

It is obvious that the sentence could be split into three where one part of the story finishes and another begins. This preserves the meaning of the original and all of its word order:

If the equipment shall go out of order, Tiny shall at its own expense have the equipment repaired by the person, firm or body corporate designated by Bigg. ~~and in~~ In the event of Tiny failing so to do then Bigg shall be entitled to take possession of the equipment and have it repaired at the cost of Tiny. ~~and during~~ During such possession and repair, the lease charges shall nevertheless accrue and be payable by Tiny to Bigg.

Then all the shalls would be replaced with 'must' or a verb in the present tense, or both. (See chapter 2.) Legalese like 'in the event of' and 'nevertheless accrue' would be translated into plain words. The rewrite could say:

If the equipment goes out of order, Tiny must at its own expense have it repaired by the person, firm or corporate body designated by Bigg. If Tiny fails to do this, Bigg may take possession of the equipment and have it repaired at Tiny's expense. During such possession and repair, Tiny must still pay the lease charges.

A long sentence is sometimes unavoidable in making a complex point which has exceptions and qualifications attached. Then, it needs to be managed well, with simple construction and plain words. This, for example, in a draft law, is not too bad:

If, in the circumstances of section 3.2 and after the expiry of a 14-day period starting on the date the agreement was entered into, the customer shows by taking some significant action that he considers the agreement to be in force, then:

(a) the customer may not give the seller notice of cancellation; and
(b) the seller may enforce the agreement against the customer, despite section 3.2.

Put people into the writing

Most legal agreements are about what people on all sides of a bargain must and must not do. It makes sense to give these people convenient names at the start of the agreement and use them throughout. So 'John Fustian of 97 Sackcloth Court, Berwick' might be identified as 'Fustian', and this term, along with 'he', 'his' and 'him', could be freely used for the sake of brevity. In standard-form agreements it is now common to define the bargain-makers by personal pronouns and this can aid clarity. In an agreement to lease a vehicle, the two sides to the bargain could be defined thus:

> **'We' means the lessor, Misfit and Snaggs Motors plc, The Garage, Sumpcity.**
>
> **'You' means the lessee, Paul Oilyhands, 7 Rag Street, Pumptown.**

The words 'we', 'you', 'our' and 'your' could then freely be used, improving every sentence in which they appear. It would mean the end of this kind of thing:

> **The lessor will register details of this lease and the conduct of the lessee's account with any licensed credit reference agency. The lessor may also disclose this and any other information supplied by the lessee to any member or associated company of the Scottish Bank plc group of companies or to any person acting on the lessee's behalf for any purpose connected with the group's business. The lessor may also use the lessee's name and address to mail the lessee about services that may be of interest to the lessee.**

Instead it would be possible to write:

> **We will register details of this lease and the conduct of your account with any licensed credit reference agency. We may disclose this and any other information you supply to any member or associated company of the Scottish Bank plc group of companies or to any person acting on your behalf for any purpose connected with the group's business. We may use your name and address to mail you about services that may be of interest to you.**

20 Basics of clear layout

Guideline: *Use clear layout to present your plain words in an easily accessible way.*

Plain English needs to be complemented by effective layout otherwise only half the job has been done. At its simplest, in a letter or short report, effective layout might mean simply using easily legible type and putting ample space between paragraphs. At its most complex, in a set of instructions or a detailed official form, effective layout might require the manipulation of hundreds of variables such as different typefaces, headings of various sizes, colours, and illustrations.

Not so long ago, layout was largely outside writers' control. Typesetting was an expensive mystery guarded by layout professionals. Now, with the advent of desk-top publishing (DTP), many writers have access to sophisticated layout tools. If these tools are not used with a reasonable degree of skill, if there is no sense of what makes a page look good, or if readers' strategies for tackling a document are ignored, the results can be dire – poor layout can negate most of the benefits of plain English.

There is no simple recipe book for good layout, but this chapter summarizes key points that you may like to bear in mind when preparing a layout or getting others to do so. Naturally, for important or high-use documents, you may need help from layout and printing professionals.

Some of the guidance is more appropriate to the highly flexible DTP environment than to word processing. The points are intended to supplement the technical guidance on using DTP that software manuals tend to cover. For simplicity I have set out the points in question-and-answer form.

What is the best way to get a feel for good layout?

Study critically the layout of some documents – books, reports, letters, forms, and your car and life insurance policies. Every line of type on every page, every ruled line, and every white space is the result of someone's conscious layout decision, for good or ill. The sum of those decisions is what gives a document its distinctive look.

Consider which documents are easy on the eye, and why? Which organize the material well, and how do they do this? If a large number of words has been squeezed into a page, what has been done (or not done) to make the result easily legible?

Consider how different layouts fulfil different purposes – to attract attention, to sell, to summarize main points, to ask questions, to act as reference material, to save paper. Consider whether the layout styles fulfil their purposes.

What is a good page size to use?

This depends on many things such as the amount of information, how the readers use the document (do they need, for instance, to carry it around with them on a building site or to follow a route?), whether it is to be stored and filed, or whether it needs to be fed through photocopiers or laser printers.

In Europe the commonest range of sizes – and therefore the most economical – are the 'A' sizes such as A3 (297x420mm), A4 (210x297mm) and A5 (148x210mm). Each of these is formed by folding in half, along the long edge, the next biggest 'A' size. All the sizes offer an infinite number of layout possibilities. Most business letterheadings are A4, and a single column of type is the obvious layout choice. But an information leaflet might have two columns to the A4 page and be double-sided; this allows a great number of words to be fitted in. An A4 page could be divided into one narrow column (perhaps for side-headings) and one wide column for the associated main text. Many leaflets are A4 folded twice on the long edge, producing a 6-panel set-up – one panel for a front cover, perhaps, and five for other information.

What are the key variables to control for high legibility?

Three important variables are type size, column width, and space between lines ('leading', pronounced 'ledding'). These interact in ways that improve or impair legibility. While it is impossible to lay down rules for how these variables should be manipulated, some guidelines may help.

Type size: Type size is measured from just above the top of the capitals to just below the bottom of letters like y and j. The distance can be measured in points. (A point is about 1/72 inch.)

In many typefaces the sizes 9-point and above will be highly legible for large areas of continuous text, though this will depend on how the other variables are handled.

Point size alone is an uncertain guide to how big the type appears to be. Here are examples of 10-point type in two typefaces, Monotype Bembo and Monotype Nimrod ('typeface' means a set of lettering in a certain design):

Example of Monotype Bembo, a widely available typeface based on a fifteenth century Italian type.

Example of Monotype Nimrod, designed in the 1970s for newspaper text setting.

Though both are 10-point, the characteristics of Nimrod make it appear bigger than Bembo. This is mainly because in Nimrod the x-height (height of the lower case *x*, *o*, *m*, *n*, etc) is great in proportion to the type size. X-height is therefore a better guide to legibility than point size. Provided the x-height is 1.5mm or more, the type will be highly legible to those with normal eyesight under good reading conditions. Documents to be read from a distance will need a bigger type size. In a reference document you might be content to use a small point size, say 7-point, for the sake of economy – but you would choose a typeface whose x-height was large enough to make the point size reasonably legible and you would use narrow columns.

Column width: For large areas of text, most layout professionals reckon that the optimum column width is 50–70 letters and spaces. This means about 8 to 12 words per line, as in the column width you are reading now.

The commonest mistake is to set small type across too wide a column, such as 170mm on an A4 page. The result could be more than 100 letters and spaces to the line unless the type is correspondingly big. And if the type is big, there will be fewer words to the page so printing or copying costs could increase.

Leading: Normally there needs to be some leading, otherwise readers tire easily and make mistakes when locating the start of successive lines. The amount of leading depends on the trade-off you make between economy and legibility. As a guide, try to ensure that the leading is about a fifth of the type size. So 12-point type might benefit from leading at 2.5 or 3 points. Generally, the wider the column, the more leading is needed.

Typefaces with a large x-height relative to their type size (like most versions of Times, Helvetica, Plantin and Palatino) tend to need the full allowance of leading; those with a smaller x-height (like Futura, Bodoni and Bembo) tend to need less.

So which is the best typeface to use for high legibility?

What works well in one set of circumstances may not work well in another. Generally, the type for body text should be quiet, simple and regular in form without the eccentricities of display typefaces such as those on the fascias of hamburger joints or on advertising hoardings. For large areas of text, it is usually better to use a serif type (that is, a type with tiny strokes or projections at the end of most of the letters). The serifs guide the eye horizontally and put light and shade on the page because the letters have thick and thin strokes. Serif types tend to look authoritative, classical and official. The text type in this book is a serif, the Monotype version of Plantin.

Highly legible serif types include Plantin, ITC Garamond, Joanna, Century Schoolbook, Palatino and Times. These are their industry names; trade names may differ for commercial or copyright reasons.

Example of a version of Joanna, set in 10-point. It has a distinctive *italic* face, and a good **bold** weight too.

Example of Century Schoolbook, again in 10-point. The *italic* face may not be very pleasing, but the **bold** type is strong and effective.

Example of Palatino, set in 10-point type. In this there is a distinctive *italic* type and a good **bold** weight.

Times is available in most DTP and word-processing software, but its narrow character width tends to make it more suitable for newspaper columns. Its universality makes it an unusual choice for any layout professionals who want to create an individual look for their client's documents. In short, it is a little boring.

The sans serif types (types without serifs) tend to be more useful as headings and in forms, catalogues and flyers, though they can look good in almost any application if handled well. Sans faces tend to be plain, unfussy and very compact, so in bold weights they make an especially strong impact. Good sans faces include Helvetica, Gill, Franklin Gothic and Frutiger.

A modern version of Gill which, like Joanna, was cut by the English artist Eric Gill. This is 10-point Gill Light, with *italics*. **This is the bold weight.**

Example of Franklin Gothic, also in 10-point but with **bold**.

Example of Frutiger in 10-point, also with the **bold weight**. The type was designed by Adrian Frutiger.

Many documents combine sans serif type for headings with serif type for body text. The opposite combination is less common but can still work well.

What about ways of emphasizing type?

Most typefaces can be used in weights such as roman (this weight), bold, italic and bold italic. A popular type like Helvetica could have as many as 28 weights. In a single document, it is usually better to use as few weights as are really necessary and to make sure there is ample difference in strength between them. (In some typefaces, especially sans faces, the italic weight is merely a sloping version of the roman and may not be noticeably different.)

Use highlighting weights sparingly – if you emphasize too much, nothing will be emphasized. If too many individual words are typed in bold, pages will look spotty or dazzling and the reader will find concentration difficult.

Most people with normal eyesight dislike reading long swathes of bold, italics, and capitals. Of these, capitals tend to be the most disruptive to reading and may seem aggressive. There is no harm in capitalizing a few words, but the usual mix of upper and lower case is the best for legibility. There is no need to set headings in capitals; generally they will look better in upper and lower case.

In typewriting, underlining was one of the few available ways of emphasizing text. The weight of rule corresponded to the weight of the type and the effect was pleasant enough in small doses. In DTP, underlining is probably the least attractive way because the line will usually <u>slice through the bottom of the type unpleasantly, like this</u>. Only do it if there's no other way.

Any hints on the use of white space?

Without wasting paper unnecessarily, you should allow generous margins and reasonable space between columns. Resist the temptation to fill white space with type. It is no tragedy if a section of a report ends halfway down a page or if the back of a leaflet has to be left blank.

If you are leaving space between paragraphs (a common alternative to indenting the first line), ensure the space adequately separates one paragraph from the next but don't allow the paragraphs to look like islands. The software will generally allow you to set up style-sheets to exert fine control over inter-paragraph space.

If headings appear in a column of text, be sure to put at least as much space above them as below them, otherwise they will appear to be floating upwards to the previous paragraph.

Is it a good idea to reverse out the type?

Reversing out means printing the type in white out of a background colour. Usually this will only be highly legible if the type is 10-point or more, in a bold weight, and if the background colour is dark. Sans serif faces tend to reverse out better than serif. Only reverse out small areas of type.

Should type be printed on coloured paper?

Not if it destroys the clarity of the type. For most purposes, there needs to be strong contrast between foreground (the type) and background. If you print dark green ink on a pale green background, you are asking for trouble, especially as about eight per cent of males are colour blind for green and red.

Is it a good idea to track or scale the type?

Tracking means adding or taking out space between letters. In this paragraph I have added extra units of tracking, with obviously unpleasant results. There comes a point when excessive tracking disrupts the readers to the extent that they start to look at individual letters instead of reading words by their shape and in sweeps of four or five words at a time.

In this paragraph, I have deducted several units of tracking, producing an equally foul outcome. If done at all, tracking needs to be sensitive and to respect the design characteristics of the typeface.

Similar points apply to the ability of some software to scale the type to compress or expand it. Don't overdo this: be sensitive to the design of the typeface.

The term 'hierarchy of headings' is often used. What does it mean?

A document may need to use several sizes or weights of heading to signify, for example, chapter headings, sub-headings and paragraph headings. This range of sizes or weights (or both) is the hierarchy of headings. In general, the strength and position of headings should reflect the job they are being asked to do. So chapter headings will usually be considerably stronger than sub-headings which will in turn be stronger than paragraph headings. This is similar to the arrangement used in this book. The hierarchy of headings should be applied consistently – readers get confused if the same signal is used with different meanings.

Should I use justified type?

Justification means inserting spaces between words (and even between letters) so that all the lines of type take up the full column width. Hyphenation at line-ends may also be needed to make this work. The main reasons for justification are economy and, according to some, greater neatness.

There is no clear-cut evidence that competent justification impairs the reading performance of literate adults. Sometimes, though, mechanical justification tends to produce rivers of white space running down the page as well as very uneven letter-spacing. If you dislike the justified type your machine produces, your software manual may offer help. For example, you should be able to adjust the program so that it only applies line-end hyphenation to words of seven letters or more and so that it bars more than two successive line-end hyphenations.

Unjustified type ('ranged left' or 'flush left') tends to produce a more relaxed, informal look. Word-spacing remains constant. The hyphenation program can be switched off if you dislike line-end hyphenation, though the drawback is a more ragged right-hand edge to the column. With unjustified type you will want to avoid breaking up the left-hand edge of the column by excessive indentation. Try to use the left-hand edge as the starting point for as much of the material as possible.

By editing or other adjustments, try to remove widows (single words forming the last line of a paragraph), especially if they appear as the first line of a page. They waste space and look unsightly. This applies equally to justified and unjustified type.

Does the use of colour help?

Cost normally prohibits the use of a second colour (or several colours) on anything except pre-printed stationery and external publications with long print-runs. But if a second colour is available, do not spatter it throughout the text. Use it mainly to help people navigate through the document, perhaps by applying it to all headings or to one level of heading. Use it also to add impact, say on a front cover. In forms, a second colour is now common (perhaps as a percentage tint) to surround white completion boxes; this helps them to stand out.

What paper should I use?

This depends on your budget and the conditions under which the document will be used. If possible, take advice from a printing professional. Heavy, gloss papers tend to be expensive and may make the type hard to read in certain lighting conditions. Thin papers create an unacceptable amount of show-through when printed on both sides. Certain weights and finishes are treated unkindly by photocopiers and laser printers. Papers with a high recycled waste content are gradually improving in quality and in their range of weights, finishes and colours.

Doesn't decent layout cost more?

The work has to be done, so it might as well be done properly. A poorly laid out form, for example, may substantially impair readers' performance and this adds to administrative costs if it has to be returned for re-completion. Government departments reckon that the cost of laying out and printing a form is as little as half of one per cent of the cost of administering it after completion. A life insurance policy might cost 20 pence (US 40c) to print, yet the first year's premiums could easily be a thousand times that. So good layout is unlikely to add significantly to costs.

To showcase most kinds of essential information, nothing more than decent competence is needed in the layout. Often the best results come from quiet, unshowy pages that the readers hardly notice. There's no need to shout.

The next four pages show layout features in two documents.

Renewal form for

Council Tax Benefit and Housing Benefit

Ref no Date issued

Name
Address

Postcode

Who this form is for
This form is for people who are getting Council Tax Benefit or Housing Benefit now. You can use it to renew one or both of these benefits.

Filling in the form
You must answer every section in the form, otherwise it will take us longer to work out your benefit.

Don't forget to sign at the end of the form. If you have a partner, he or she must sign the form as well. (Partner means someone you are married to, or someone you live with as if you are married to them.)

We will treat anything you tell us as confidential.

After you have filled in the form
When you have filled in the form, send it back to us straight away. Then, if you are still allowed benefit, we will carry on paying it to you without a break.

If you do not send us the form, or there is a delay, you may lose benefit. This usually means you will have to pay more Council Tax or rent.

After we have worked out your benefit
After working out your benefit we will send you a letter saying how much you will get. If your Council Tax Benefit has changed, we will also send you a bill for the new amount.

If you cannot get benefit
If we cannot renew your benefit, we will write and tell you why. The letter will tell you how to appeal if you think our decision is wrong.

please turn over

Page from a form, showing a hierarchy of heads and vertical spaces

Paragraph heads are nearer to the text below them than above them, while spaces between paragraphs are greater than those between sub-paragraphs. Line length is about 70 characters. The background is 10 per cent black; in practical use a pale pastel would be better. The column is offset slightly to the right – more energetic than centring.

Part A About you

1 What is your surname?

2 What are your first names?

3 What is your title? *(for example Mr, Mrs or Miss)*

4 What is your date of birth?

/ /

5 What is your National Insurance number? *(if you know it)*

6 What is your daytime telephone number, if you have one? *(You do not need to tell us, but it might speed up your claim if we need to contact you.)*

Part B About your partner

1 Do you have a partner who normally lives with you?

No ☐ Go to **Part C**

Yes ☐

2 What is your partner's surname?

3 What are your partner's first names?

4 What is your partner's title? *(for example Mr, Mrs or Miss)*

5 What is your partner's date of birth?

/ /

6 What is your partner's National Insurance number? *(if you know it)*

7 Is your partner registered blind?

No ☐

Yes ☐

2

Second page of the form

This continues the layout themes of the first page and reveals another reason for using a wide left-hand margin – it makes room for navigational markers like 'Part A'. Question numbers hang outside the wide column in their own space; they are also picked out in bold. At full size, the answer spaces are about 8mm deep, adequate for normal handwriting. It's a very spacious layout.

INTRODUCTION

We're here to help

Lots of people are in debt these days. Sometimes it's their own fault, but usually it's not.

National Debtline is the national telephone helpline for people with debt problems. We give expert advice over the telephone and send every caller this information pack free of charge. The service is free, confidential and independent.

What this leaflet covers

This leaflet gives you good advice on tackling your debts. It shows:

- How to work out your **Personal Budget** (a Personal Budget sheet comes with the leaflet). You can use this to explain your money problems to the people you owe money to (your 'creditors'). Everyone wants you to pay *their* bill, but they don't stop to realise that you have other problems. Seeing the Personal Budget may stop them all chasing you for money at the same time.
- How to decide which debts to deal with first—your **priority debts.**
- How to make reasonable offers to repay your creditors.
- How to cope with court procedures.

But if you'd like to discuss anything in more detail, please phone us—the number and opening hours are on the front.

Page from a debt advice booklet

This introduces the booklet's principal topics with two main headings and a bullet-point list. As this is the first left-hand page of the booklet, there is no sense in filling it with text – right-hand pages are likely to be examined more thoroughly. The page benefits from a jaunty graphic which emphasizes that advice is only a phone call away.

Mortgage arrears

You may have both first and second mortgages.

The **first mortgage** is the loan you took out to buy your home.

The **second mortgage**, also known as a *secured loan, second charge* or sometimes a *consolidated loan plan*, is a separate loan which is secured on your home.

Check all your loan agreements to see if they are 'unsecured' or 'secured' on your home. If 'secured' treat them as priority debts because lenders can ask the court for possession of your home if you cannot pay your monthly instalments. The property can then be sold to pay off your debt.

The legal term for the company or building society who gave you your mortgage is a 'mortgagee'. In this leaflet we call them 'lenders'.

Contact your lenders if you have problems

It's never too early or too late.

You may not be in arrears yet or your lenders may have started court action. Whatever the situation, **do not delay** contacting your lenders. Get in touch as soon as possible by writing, phoning or making an appointment to see them.

If the local office is unhelpful or difficult, contact their head office and try to reach an agreement with them.

If you have not paid the mortgage for a number of months it's very important that you start paying what you can, even if this is not the full monthly payment.

Arranging to clear the arrears

Warning

Don't be tempted to take out an extra loan to repay your mortgage arrears. Often these are very expensive and could put your home at greater risk.

You will usually have to offer an extra monthly payment to clear the arrears. The lenders will normally ask for the arrears to be cleared over 12 to 24 months. But even longer periods can be agreed in some circumstances. If you cannot manage to clear the arrears within the time the lenders want, start paying the amount you have offered anyway. Explain why you can't pay the amount they have asked for, particularly if there are circumstances such as a long illness, death of a partner, marriage breakdown or unemployment.

If the value of your home is far more than the total mortgage, tell the lenders. The more your house is worth, the less risk your lenders are taking.

If your home is not worth as much as the mortgage, see the section on page 10.

Sometimes your lenders may be able to accept another arrangement. We describe some of them below.

Adding the arrears to your mortgage

This is called 'capitalising' the arrears.

Normally you can only do this:

- on first mortgage arrears; and
- if the value of your property is a lot more than the total amount of your mortgage.

It works like this: the amount of the arrears is added to the total mortgage. The monthly repayments are increased to take account of this. So the arrears are spread over the remaining years of the mortgage term.

Your lenders may be more likely to agree to this if you have already kept to a payment agreement for some months—it shows you are able to pay.

Changing to a repayment mortgage

If you have an endowment mortgage you may be able to change this to a repayment mortgage. Endowment mortgages include an insurance policy and if you have had this policy for a few years it may have a surrender value. The surrender value is the amount of cash the policy is worth if you cancel it.

Ask your lenders about this and get advice on whether:

- it is a good idea to cash in your endowment policy;
- changing to a repayment mortgage will reduce your monthly payments.

If you cancel your endowment policy, ask your lenders about a mortgage protection policy. This would cover the payments if you died.

If you do change to a repayment mortgage you should also ask your lenders to extend the mortgage term to 25 years. (See below.)

Increasing the mortgage term

Most mortgages are spread over 25 years — this is called the mortgage term. If you have already lived in your home for several years, you could ask your lenders to extend the term back to 25 or even 30 years. This could cut the regular monthly payment so that you can afford to pay something towards the arrears each month. If you have an endowment mortgage, this may be more difficult. Ask your lenders.

Same booklet, later right-hand page

This two-column layout enables plenty of text to be included in an A4 page, while still remaining easy on the eye. In the published version, headings are printed in green with the rest of the type in black. The column width allows about 45 letters and spaces to the line.

Sources and notes

General sources

Butcher J *Copy-editing* (Cambridge 1992).

Eagleson RD *Writing in Plain English* (Canberra 1990).

Felker DB and others *Guidelines for Document Designers* (American Institutes for Research 1981).

Gowers E *The Complete Plain Words* (London 1986).

Greenbaum S *An Introduction to English Grammar* (Harlow 1991).

Hart's Rules for Compositors and Readers (Oxford 1983).

Howard G *The Good English Guide* (London 1993).

Miller C and Swift K *The Handbook of Non-Sexist Writing* (London 1989).

Turk C and Kirkman J *Effective Writing* (London 1982).

Williams JM *Style* (Glenview, Illinois 1981).

Legal language

Adler M *Clarity for Lawyers* (London 1990).

Asprey MM *Plain Language for Lawyers* (Sydney 1991).

Charrow VR and Erhardt MK *Clear and Effective Legal Writing* (Boston 1986).

Cutts M *Lucid Law* (Stockport 1994).

Garner BA *Dictionary of Modern Legal Usage* (Oxford 1987).

Mellinkoff D *The Language of the Law* (Boston 1963).

Wydick R *Plain English for Lawyers* (Durham, N Carolina 1985).

Specific sources and notes

Starting points

page 4 For the development of plain English, see the *Oxford Companion to the English Language*, edited by Tom McArthur (Oxford 1992).

page 5 The remarks of Coode and Bentham are quoted in *Writing Rules: Structure and Style*, an unpublished paper given by David C Elliott to the International Conference on Legal Language at the Aarhus School of Business, 1994.

page 7 The studies on jury instructions are quoted in Joe Kimble's paper *Answering the critics of plain language* (unpublished, Thomas Cooley Law School 1994). The primary sources are Charrow RP and Charrow VR *Making Legal Language Understandable: A Psychological Study of Jury Instructions*, 79 Columbia Law Review 1306, 1370 (1979); and Benson RW *The End of Legalese: The Game Is Over*, 13 New York University Review of Law & Social Change 519, 546 (1984–85).

page 8 Work on the Clearer Timeshare Act is published in Cutts, above. This and other efforts to clarify statutes are explored in *The Law-Making Process* by Michael Zander (London 1994).

chapter 1

page 12 The analysis of American sentence length is included in *Computer Analysis of Present-Day American English* by Kucera H and Francis WN (Providence, Rhode Island 1967).

chapter 2

page 26 The research study among scientists is described in Turk and Kirkman, above.

page 31 The Survey of English Usage is managed by the English Department, University College London.

chapter 3

page 40 Strunk is quoted in *The Elements of Style*, Strunk W and White EB (New York 1979).

chapter 4

page 48 Orwell's prescriptions are quoted under his entry in the *Oxford Companion* (above) and in his essay *Politics and the English Language*, included in *Shooting the Elephant and Other Essays* (London 1950).

page 54 Among those who recommend the use of 'I' and 'we' are Turk and Kirkman (above); Booth V *Writing a Scientific Paper* (London and Colchester, The Biochemical Society 1981); Sandman and others *Scientific and Technical Writing* (CBS College Publishing 1985); and Barrass R *Scientists Must Write* (London 1978).

page 55 Information on the passive percentage is drawn from a manual accompanying the software program *StyleWriter – the Plain English Editor* (Editor Software Pty Ltd 1994).

chapter 11

page 81 The role of punctuation in showing the construction of the sentence is described in GV Carey's *Mind the Stop* (London 1958).

page 83 My dislike of the 'nameless thing' is shared by *Hart's Rules*, above.

chapter 12

page 95 Nesfield's book, long out of print, was published by Macmillan, London.

chapter 14

page 103 The core statement is a variation of the core sentence described in *Write To Win* by Thomas McKeown (Clear Communications Press, North Vancouver 1987).

page 106 For the section on different approaches to planning, I have drawn upon *Writing Strategies and Writers' Tools* by Daniel Chandler (*English Today* 34, Vol 9, 2 1994).

chapter 16

page 123 The use of algorithms and other alternatives to continuous prose is thoroughly described in James Hartley's *Designing Instructional Text* (London 1991).

chapter 19

page 140 The chairman of Clarity, Mark Adler, was writing in the *Information Design Journal* 7/2 (1993): *Why Do Lawyers Talk Funny?*

Index

acronym 86, 87
adjective 99
adverb 99
algorithms 121–3
'and' 95
apostrophe 89–92
Austen, Jane 94

Barnes, William 5
Basic English 5–6
Bentham, Jeremy 5, 140
brackets 85, 86
bubble diagram 102–3
'but' 13–14, 94–5

capital letter 65–6, 86–7, 152
Chaucer, Geoffrey 4
chronological order 111
clause 100
co-ordinator 31, 100
colon 81, 83
colour 153, 154–5, 156, 159
comma 81–3
connectors 13, 95
contractions 17, 100
Coode, George 5
core statement 103–4
correspondent's order 115
cross-reference 69–70

dash 83, 85–6
definition of plain English 3–4
desk-top publishing 148
doer 49–52, 54, 100, 101

Edward VI, King of England 4
ellipsis 92
exclamation marks 93

focus group 3
  ratings 18, 23, 47, 52, 121, 130–1, 136, 145
full stop 12–13, 17, 65–6, 81

gobbledygook 19
Gowers, Sir Ernest 6, 19
grammar 63–4, 99–101

'he/she' 72–3
headings 76–7, 87, 136–7, 148, 152–3
  hierarchy of 154–6
history of plain English 4–7
honesty, plain English and 4
horizontal plan 104–5
hyphen 88–9, 154

illustrations 137–8, 148, 158
imperatives 100, 133–6
infinitives 63–4, 100
  splitting 96–7
inkhorn terms 5
instructions, writing 132–9

journalists 80
justification 154

lawyers 21, 48, 67, 140–3
layout 18, 118–23, 148–59
legal documents 21–2, 140–7
legibility 149–52

letters 75–9
lists 16, 102, 119, 158
  punctuation 82–5
  vertical 15, 17–18, 61–6

management of colleagues' writing 124–31
margins 152, 157
memos 113–15

negative to positive 67–8
Nesfield, J. C. 95
nominalization 56–9, 100
notwithstanding 21
noun 56, 100–1
numbering:
  lists 66
  paragraphs 115, 135–6

Ogden, C. K. 5
'only' 68
openings and closings of letters 78–9
Orwell, George 48, 95

page size 149
paper 155
paragraph:
  definition of 101
  layout 148, 153, 156
  length 97
  numbering 115, 135–6
  presentation of information 119, 134
parallel structure in lists 62–3
participle 101
  past 50–1
planning 17, 70, 102–7
platform statement 62–3, 65–6
plural 73, 101
pomposity 19, 26, 43–4, 142–3
position, problem, possibilities, proposal, packaging structure 114
Potter, Beatrix 96
preposition 96, 101
problem–cause–solution structure 111
pronoun 101
punctuation:
  accuracy 80–93
  of address 79
  of lists 64–6

Quakers 5
question-and-answer structure 111–12, 120–1
question mark 93
quotation marks 92–3

readability formula 98
reader:
  centred structure 108–17
  instructions for 133
redundant words 15–16, 40–6
  see also verbiage
repetition 14, 41
reports, writing 53–4, 115–17
reversing out 153
Richards, I. A. 5

sans serif type 151–2, 153
semicolon 65–6, 81, 83–4
sentence:
  definition of 101
  half/full 75
  length 2, 11–18, 145–6
sexist usage 71–4
situation, complication, resolution, action, politeness structure 112–13
situation, objective, appraisal, proposal structure 113
speech, writing as 97
splitting:
  infinitives 96–7
  sentences 12–14
structure 108–17
Strunk, Professor William 40
survey of English usage 31, 37
synonym 22, 26

table layout 118–21
teamwork 125–7
tense 101
testing instructions 138
thinking and planning 102
tie-breaks 51
'to be' 50, 57
'to have' 50, 57–8
tracking 153
triangle structure 108–10
Tyndale, William 4
typeface 148–55

verb:
  active/passive 48–55, 63
  definition of 101
  feeble 44–5
  letter writing 76
  vigorous 56–60
verbiage 15–17, 20, 44
voice, active/passive 48–55, 58–9, 63, 101, 134

warnings 137
'we', use of 53–4
words:
  archaic 20, 22, 77–8
  foreign 37–9
  legal 141–6
  obsolete 24
  plain 19–39
  technical 2, 21
  unusual 19–22
writing:
  managing colleagues' 124–31
  myths 94–8
  scientific 26, 54, 115
  technical 21, 54, 115
  tight 40–7